W9-AFE-809

UNDERSTANDING

Environmental Policy

UNDERSTANDING Environmental Policy

Steven Cohen

COLUMBIA UNIVERSITY PRESS

New York

Columbia University Press
Publishers Since 1893
New York Chichester, West Sussex

Library of Congress Cataloging-in-Publication Data
Cohen, Steven, 1953 –
Understanding environmental policy / Steven Cohen.
p. cm.
Includes bibliographical references and index.
ISBN 0-231-13536-X (acid-free paper) —
ISBN 0-231-13537-8 (pbk. : acid-free paper) —
ISBN 0-231-50962-6 (electronic)
1. Environmental policy. 2. Environmental management. I. Title.
GE170. C62 2006
363. 7'0561—DC22 2005034279

♾

Columbia University Press books are printed on permanent and durable acid-free paper.
Printed in the United States of America
c 10 9 8 7 6 5 4 3 2 1
p 10 9 8 7 6 5 4 3 2 1

To Gabriella Rose and Ariel Mariah and a future to cherish . . . .

Contents

Preface

Public policy analysts know that one cannot begin to solve a problem until it is understood. This includes measurement of the problem's dimensions and proposed solutions. One cannot manage something without measurement. Without measures, one cannot tell if the management actions taken are making matters better or worse. Without a deep conceptual understanding of the problem, the process of developing measures of its elements cannot begin. Think of the modern economy. We collect, analyze, and report a vast array of indicators, ranging from employment to consumer confidence to gross domestic product. Even then, our management of the economy is far from 100 percent successful. The economy is the collective actions of human beings within a system designed by human beings. The environment is exponentially more complex than the economy. We have the collective actions of humans in a system bounded by rules of ecology, physics, chemistry, and biology. This integration of human-made systems within natural systems is extraordinarily intricate, and we are only beginning to understand its dimensions. It is the hope of this author that the framework offered in this book will help to navigate policy makers and analysts through this intricate mesh of natural, political, social, and scientific systems.

In our very human arrogance we somehow think that we are further along in understanding environmental problems than we actually are. The framework proposed here is not intended as the last word on the subject—far from it. It is only a small step in a conversation that began in earnest over the last half-century. For me, it started in 1975, in Lester Milbrath's seminar in Environmental Politics at the State University of New York at Buffalo. With his teaching assistant, Shel Kamienicki, providing much needed additional guidance, Milbrath introduced me to such pivotal works as *The Limits to Growth* by the Club of Rome (Donella H. Meadows, Dennis L. Meadows, Jorgen Randers, and William W. Behrens II) and its less pessimistic successor *Mankind at the Turning Point* by Mihajlo Mesarovic and Eduard Pestel. At that time the issue of the planet's health was new to me and seemed incredibly important. I was also exposed in that seminar to scholars who explored the deeper meaning of the environmental issue, such as Robert Hielbroner, author of *An Inquiry into the Human Prospect,* and its postscript, "What Has Posterity Ever Done for Me?"

I began to explore the environment as a policy issue nearly three decades ago. In almost thirty years of studying, teaching, and working in and around government, I still feel incredibly ignorant about environmental policy. Despite my sense of inadequacy in the face of this enormous issue, I think that time is running short. We must begin to codify what we know and develop a true profession of environmental management. In the last three years I have helped build graduate programs at Columbia University in environmental science, policy and management, climate and society, and sustainable development. Perhaps this new academic arena provides some evidence that our exploration and base of knowledge about environmental issues is expanding.

However, I once read a story that makes for an interesting analogy to our present position. There came a point at the start of the twentieth century when a visit to a physician's office actually increased rather than decreased the probability of recovery and cure. Prior to that point one was better off without treatment. In this field I have days when I think we have not even opened our office for professional practice. We may not even be at the stage of bleeding our patients to affect a cure. We have managed to reduce pollution, but are we introducing more toxics than we are cleaning? We have created more wealth and material consumption than our ancestors could have imagined, but is it sustainable? And what about the billion people on the planet living in dire and abject poverty?

Still, what choice do we have but to try to address the issue of protecting the environment? Hope for the future is hardwired into the culture I was raised in, and I believe that some knowledge and understanding is better than none. A little knowledge *may* be a dangerous thing, but no knowledge is far worse. The environment may be a complicated policy issue, but we must do what we can to increase our understanding of it and improve its current condition. This book represents my best thinking on its dimensions.

I have been interested in many other issues professionally, such as public management, organizational innovation, homelessness, welfare-to-work, public ethics, and others, and yet I constantly return to issues of environmental policy. Meeting the basic human biological necessities of breathing, drinking, and eating somehow always transcends other issues in importance. In one of President Kennedy's most enlightened speeches, he found that the danger of nuclear war, at its most fundamental level, is an environmental issue. More than forty years later, I still find his thoughts moving and an appropriate coda to this introduction:

> For, in the final analysis, our most basic common link is that we all inhabit this small planet. We all breathe the same air. We all cherish our children's future. And we are all mortal.
>
> John Kennedy, American University, June 10, 1963

Acknowledgments

This book represents a journey from Brooklyn, New York, where I was raised, to Buffalo, New York, where I first learned about environmental policy, to Washington, D.C., where I worked for the U.S. Environmental Protection Agency (EPA) as an employee and a consultant, to Morningside Heights in New York City where I have worked for Columbia University's School of International and Public Affairs since 1981 and for the Earth Institute since 2002. It is a journey of exciting ideas, new policy and management concepts, amazing people and wonderful work, and three decades of learning.

Along the way I have worked with, and for, some wonderful people and learned about public policy, politics, public management, and environmental policy. In Buffalo, my teachers included Lester Milbrath, Sheldon Kamieniecki, Harry Klodowski, Jon Czarnecki, Richard Tobin, Frank De Giovanni, Marilyn Hoskin, Tony Khater, Fred Snell, and the incomparable Marc Tipermas. In Washington, I learned from a great number of people including Alan Altshuler, Ronald Brand, Bonnie Casper, Michael Cook, Michael Farber, Mary Ann Froelich, Thomas Ingersoll, James Janis, Tony Khater (again), Paul Light, Sylvia Lowrance, Andrew Mank, Sammy Ng, Robert O'Connor, Charles Smith, and once again, as my boss in D.C., Marc Tipermas. This essay grew out of a lecture I gave at the Universidad Externado de Colombia in 1998, and I thank my good friend Professor Mauricio Perez Salazar for his encouragement and hospitality.

Here in New York City the number of people I have learned from is too large to list, but I will try to acknowledge some: my coauthor on most of my work in public management, and my good friend, William Eimicke; Demetrios (Jim) Caraley, the long-time editor of *Political Science Quarterly*, the founding director of Columbia's MPA Program, and professor of political science; and Michael Crow, now president of Arizona State University, founder of the Earth Institute, and the person who challenged me to get back into environmental policy. I wish to thank four brilliant senior staff who did such a good job running things and telling me what to do that I was actually able to think and write: Dorothy Chambers, Nancy Degnan, Barbara Gombach, and Louise Rosen. I am grateful for the support and advice I have received from the deans at the School of International and Public Affairs (SIPA) I've worked for: Harvey Picker, Al Stepan, John Ruggie, and Lisa Anderson. I also thank my inspirational and brilliant boss at the Earth Institute, Jeffrey Sachs, and my colleagues on the management team at the Earth Institute: Ji Mi Choi, Mary Ellen Gallagher, Alison Gilmore, Katie Mastriani, Michael Klompus, Gordon McCord, Ryan Meyer, John Mutter, Dan Nienhauser, Barbara Noseworthy, and Mary Tobin. I am also grateful for the hard work of two key staff members in my office, Yana Chervona and Kelly Quirk.

I owe a great debt to the public policy faculty I worked with over the years, including Howard Apsan, the late Robert W. Bailey, Tom Banker, Larry Brown, Steve Cameron, Robert Cook, Mayor David Dinkins, David Downie, Mel Dubnick, Ester Fuchs, Michael Gelobter, Lewis Gilbert, Adela Gondek, Mark Gordon, Tanya Heikkila, Robert Lieberman, David Maurrasse, Eileen McGinnis, Kathleen Molz, Richard Nelson, Ralph Nunez, Dan O'Flaherty, Alex Pfaff, Blaine Pope, Anne Reisinger, Elliot Sclar, Andrea Schmitz, Glenn Sheriff , Fred Thompson, Sara Tjossem, Jacob Ukeles, Bogdan Vasi, Harold Watts, Gary Weiskopf, and Paula Wilson. I also thank several environmental scientists I worked closely with and learned from: Mark Cane, Marina Cords, Patrick Louchouarn, Don Melnick, Bob Pollack, Stephanie Pfirman, and Nickolas Themelis. I am also grateful for the lessons I learned from two of EPA's senior managers in Region II: Kathy Callahan and George Pavlou.

I thank my research assistant during the summer of 2004, Kate Brash, for the careful and creative work she did in helping me produce this book. Kate, a graduate of Columbia's MPA in Environmental Science and Policy (2004), is now a senior staff member of the Earth Institute. I also thank my research assistants, Naomi Zuk, Amy Wiedemann, and Laura Zaks, for helping me revise and edit the manuscript for publication. Naomi and Laura are graduates of Columbia's School of International and Public Affairs, and Amy will be, too, at the time you read this book. I am grateful to Gina LeVeque, a student in SIPA's Master of International Affairs Program for her superb work on the index. I also thank Rita Bernhard for the thoughtful and thoroughly professional work she did to copyedit this book.

I continue to be grateful for the love and support of my family: my wonderful wife, Donna Fishman; my children, Gabriella Rose and Ariel Mariah; my parents, Marvin and Shirley; my brother, Robby; and my sisters, Judith and Myra.

Part I

Developing a Framework

Chapter 1

Understanding Environmental Policy

Differing Perspectives on Environmental Policy

Environmental policy is a complex and multidimensional issue. As Harold Seidman observed in *Politics, Position, and Power: The Dynamics of Federal Organization*, "Where you stand depends on where you sit." Put another way, one's position in an organization influences one's stance and perspective on the issues encountered. Similarly, one's take on an environmental issue or the overall issue of environmental protection varies according to one's place in society and the nature of one's professional training.

For example, to a business manager, the environmental issue is a set of rules one needs to understand in order to stay out of trouble. For the most part, environmental policy is a nuisance or at least an impediment to profit. Perhaps someday business managers will see it as a set of conditions that facilitate rather than impede the accumulation of wealth. For now, most business practitioners see a conflict between environmental protection and economic development, although this view of a trade-off is false. To an engineer, the environmental problem is essentially physical and subject to solution through the application of technology. Engineers tend to focus on pollution control, pollution prevention (through changes in manufacturing processes or end-of-pipeline controls), and other technological fixes. Lawyers view the environment as an issue of property rights and contracts, and the regulations needed to protect them. Economists perceive the environment as a set of market failures resulting from problems of consumption or production. They search for market-driven alternatives to regulation. Political scientists view environmental policy as a political concern. To them, it is a problem generated by conflicting interests. Finally, for philosophers, the environment is an issue of values and differing worldviews.

The environment is subject to explanation and understanding through all these disciplines and approaches. It is, in fact, a composite of the elements identified by the various disciplines and societal positions, and likely has dimensions where the disciplines and social perspectives intersect. The difficulty is that each view tends to oversimplify environmental problems, contending with only one facet of the situation. Although such problems are multidimensional, different types of environmental issues are weighted toward different conceptual orientations. One view may explain a greater or lesser share of the problem than another. For example, the problem of leaking underground storage tanks is not a technical issue, because we know how to prevent leaks; the technology need not be developed anew. Nor is it a problem of economics, for the materials leaking from the tanks are products that have a recognized value. Market mechanisms would indicate that the costs of preventing leaks are far lower than the benefits, not only to society as a whole but also to most tank owners. However, that many underground storage tanks leak is primarily a management problem: many businesses that own tanks simply do not have the organizational capacity to prevent them from leaking.

Developing a Framework to Help Understand Environmental Issues

This chapter is intended to contribute to a conversation about the environmental problem in general, as well as to certain areas in greater detail. The environmental problem can be defined as the set of interconnected issues that determine the sustainability of the planet earth for continued human habitation under conditions that promote our material, social, and spiritual well-being. In chapter 2 I develop a framework for understanding the dimensions of the environmental problem and solutions proposed to address these problems. The framework allows us to deconstruct particular environmental issues and programs to understand their causes and effects. It examines environmental issues as a multifaceted equation encompassing a variety of factors including values, politics, technology and science, public policy design, economics, and organizational management. Each aspect of the framework illuminates a specific feature of the environmental issue and at the same time clarifies all the environmental issues examined here. Each separate issue, however, tends to find its main source of explanation in a single factor.

Applying the Framework to a Set of Environmental Issues

Any number of issues could have been selected to apply the framework I develop in chapter 2. I selected issues I have experience analyzing, of course, but also those that are significant in terms of policy and that vary depending on the level of government most involved. Once the framework is presented in chapter 2, the remaining chapters make use of that framework to address the environmental issues intrinsic to this book: chapter 3 describes and analyzes the garbage crisis in New York City; chapter 4 addresses the problem of leaking underground storage tanks; chapter 5 applies the framework to the cleanup of toxic waste sites; chapter 6 details and characterizes the issue of global climate change; chapter 7 compares the issues and discusses both the strengths and limitations of the framework, and also identifies possible modifications; and chapter 8 offers suggestions for improving environmental policy.

The issue of waste management in New York City, the topic of chapter 3, may well represent the future of waste management in the United States. In nations like Japan approximately 70 percent of the waste is incinerated and used as fuel to generate electricity. Most parts of the United States still have a great deal of relatively inexpensive land available for landfills. Even New York City, until 2001, had enough land to dump most of its garbage in landfills, although today the city must export all its garbage to out-of-state landfills and incinerators. Proposals to build incinerators in each of New York's five boroughs have come within days of acceptance but have always been rejected in the end.

The city's current method of waste disposal is the most expensive, environmentally damaging option one could imagine. Waste-to-energy plants would be more cost-effective and less polluting than the current system, but the politics of situating these plants causes the city government to pursue a policy of waste export. Why is it so difficult for New York City to formulate an effective waste plan, when cities like Tokyo and Barcelona manage their own waste? Questions such as these are addressed in chapter 3.

In the United States waste management is largely administered by local governments. Although solid waste management is regulated at the federal level, municipal solid waste is usually an issue of local politics and policy. Leaking underground storage tanks involves all levels of government, but because most of the leakage from these tanks is gasoline, and because the majority of companies that make gasoline are multinational corporations, policy regulating these tanks has tended to be national in scope.

Why leaking underground tanks became a political issue is an interesting question. The material leaking from the tanks and polluting the environment has not yet been used or sold to a consumer. Because it is a product with economic value, however, the owner of the gasoline is highly motivated to keep it from leaking in order to sell it. Surely we have the technology to make tanks with a low probability of leaking. No political lobby is arguing that we must preserve our leaking underground tanks. Why, then, does this problem persist? More than twenty years after we began regulating underground tanks, thousands leak every year. So how did this problem emerge as an environmental issue and a public policy concern? Why does the problem persist, and what can we do to address and solve it?

Leaking underground gasoline tanks remind us of the fragility of ecosystems and the ability of humans to inadvertently destroy nature. Although some environmental damage is a direct and unavoidable by-product of industrial production, it is human error that causes tanks to leak. Leaking tanks do not enhance the profits of oil companies. There is no inevitable trade-off between environmental protection and the generation of wealth; mistakes cause pollution from underground tanks. Our own carelessness or the very human tendency to err is to blame. One can probe deeper, of course, into why tanks leak. If we ask, for example, how the tanks got there in the first place, we need to examine the factors that generated suburban sprawl. If we ask why the fuel for our transportation is so toxic, we need to look at the development of the internal combustion engine and the technical, economic, and value factors that led to its dominance. Chapter 4 uses the framework discussed in this book to help us understand the problem of leaking tanks and the policies designed to address that problem.

The problem of toxic waste cleanup is the next issue we examine. In the twentieth century we developed an industrial economy in the United States that applied a variety of human-made chemical technologies to create a dazzling array of products, ranging from nylon to make more durable sweaters to plastics that allow me to view the screen on the laptop to write this book. With the same impulse that drove us to landfill our garbage, we assumed that once the wastes we generated from these production processes were buried underground, they were gone forever.

Few of us realized the toxicity of this waste, and even fewer how the toxic materials were transported through the ground, water, and air. Today engineers have developed a field called industrial ecology with the goal of creating products without generating waste. In the mid-twentieth century engineers paid little attention to the generation of waste when designing production, thinking perhaps that "you can't make an omelet without breaking some eggs." The rush to produce could not be delayed because of concerns about waste. Indeed, until W. E. Deming (1986) demonstrated that higher-quality products could be made with less wasted time, materials, and labor, most operations engineers and managers took few pains to reduce waste.

On learning about toxic waste contamination in the late 1970s and early 1980s, efforts were made to clean up the areas that had been damaged and prevent new waste sites from being created. But few recognized just how much damage had been done or how expensive and difficult, if not impossible, such cleanups might be. How did we create such a lethal landscape, and how did the issue reach the policy agenda? How was it defined, and what made us believe that toxic waste sites could even be cleaned up? What did toxic waste teach us about environmental problem solving?

It was largely the development of the Superfund program that led us to define environmental protection as a policy area concerned with human health. Environmental policy no longer focused exclusively on preserving mountain streams and protecting wildlife but now was also concerned about keeping poisons out of suburban basements and backyards. What was the social, political, and economic impact of this change, and how did it come about? Chapter 5 attempts to answer these questions by deconstructing the toxic waste problem into its component parts.

The final issue examined in this book is that of global climate change, in many respects the most complex environmental problem confronting us. The earth's biosphere is an extremely complicated system and one that science does not fully understand. While we know that the planet has experienced nonhuman-induced climate changes over time, we do not entirely comprehend those natural cycles and so, in the 1970s and 1980s, we were uncertain whether some changes we were noticing were natural or human-made. By the turn of the twenty-first century, scientific hesitancy was fading, and it became clear that the carbon dioxide emissions from our use of fossil fuels was causing global warming.

Pollution in one part of the planet can affect another place far away. Air pollution from power plants in the Midwest, for example, can impair the air quality in New York City. Still, the degree of global impact from air pollution is limited. The air in my home city of New York does not appear to be polluted from impurities in Mexico City or Hong Kong; for that we need to thank Cincinnati and parts of Illinois. Climate change is the first environmental issue we know about that is truly global in character. Carbon dioxide emitted from an SUV in suburban Houston contributes to raising temperatures across the planet. Carbon dioxide, of course, is not our only global environmental problem, but it is the first one that scientists have been able to use to educate the public.

While the problem of toxic waste can be addressed at the local level, a local approach to climate change can only work if it is part of a coordinated worldwide effort. The need for global action poses a challenge to our international system of diplomacy. Historically the nation-state derived from the need for security and the ability of that form of government to provide that security. Technology, however, appears to have threatened the viability of the nation-state in at least three ways. The first was the development of the atomic bomb. Nuclear proliferation challenges the capacity of the nation-state to provide security. The second threat came with the development of the Internet, containerized and air shipping, microcomputers, and satellite communication. The very technology that made the global economy possible has impaired national economic self-determination. The third threat comes from the generation of energy for electricity and transport, which has resulted in excessive releases of carbon dioxide. The result may be other forms of global ecological damage, and surely a reduction in the effectiveness of national environmental policy.

Chapter 6 analyzes the origin and impact of climate change, an impact that is more difficult to project than many other environmental issues. While we can track the introduction of a chemical pollutant into the environment and measure its effects on human and ecological health, the influence of climatic change will vary in ways we cannot predict and will not resemble the patterns of impacts we have seen with other environmental issues. Some areas may actually benefit from improved agricultural productivity that results from warmer weather and increased rainfall; others could suffer from a rise in sea level; and still other areas could be damaged by drought.

Toward an Interdisciplinary Understanding of Environmental Policy

The goal of the framework for discussion in this book is to engage in a conversation across disciplines. In the effort to understand environmental policy, one must learn some science, engineering, economics, political science, organizational management, and even other branches of learning. The power and dominance of individual academic disciplines, however, makes it difficult for an interdisciplinary conversation to take place with the same rigor and intensity that occurs within disciplines. I invite those with expertise in a particular discipline to critique the framework proposed in this book and improve it, with the aim of developing a more powerful set of tools for understanding the complex issue of environmental policy, a theme I return to in the concluding chapter.

Chapter 2

A Framework for Understanding Environmental Policy

Environmental problems cross the boundaries of sovereign states and affect natural systems worldwide, as in the case of global climate changes. The environmental problem is multidimensional, linked to the inescapable fact that human beings are biological entities that depend on a limited number of resources for survival. As the earth's population continues to grow, so does the stress on finite natural systems and resources. Yet our ability to use information and technology to expand the planet's carrying capacity also continues to grow.

This book is a brief exploration into the fundamental issues of environmental policy. It presents and applies a rough framework for a multidimensional analysis of environmental issues. The cases analyzed range from the disposal of city garbage to the complex scientific controversy of global climate change. The cases vary by technical complexity, level of government involvement, and scope of potential impact. The cases are selected to illustrate the usefulness of examining them from these vantage points. Other cases could easily be selected. The framework itself is a work in progress. It provides a method for looking at environmental issues from more than one perspective. By applying the framework to specific cases, a practitioner, student, or analyst is able to observe aspects of the issue that might otherwise be easily ignored.

For purposes of this analysis, an environmental problem is conceptualized as follows:

- *An issue of values.* What type of ecosphere do we wish to live in, and how does our lifestyle impact that ecosphere? To what extent do environmental problems and the policy approaches we take reflect the way that we value ecosystems and the worth we place on material consumption?

- *A political issue.* Which political processes can best maintain environmental quality, and what are the political dimensions of this environmental problem? How has the political system defined this problem and set the boundaries for its potential solution?
- *A technology and science issue.* Can science and technology solve environmental problems as quickly as they create them? Do we have the science in place to truly understand the causes and effects of this environmental problem? Does the technology exist to solve the environmental problem or mitigate its impacts?
- *A policy design and economic issue.* What public policies are needed to reduce environmentally damaging behaviors? How can corporate and private behavior be influenced? What mix of incentives and disincentives seem most effective? What economic factors have caused pollution and stimulated particular forms of environmental policy? Economic forces are a major influence on the development of environmental problems and the shape of environmental policy. In this framework we view these economic forces as part of the more general issue of policy design. While most of the causes and effects of policy are economic, some relate to other factors such as security and political power.
- *A management issue.* Which administrative and organizational arrangements have proven most effective at protecting the environment? Do we have the organizational capacity in place to solve the environmental problem?

This multifaceted framework is delineated as an explicit corrective to analysts who narrowly focus on only one or two dimensions of an environmental problem. The paragraphs that follow discuss policy and management approaches typically used to "solve" environmental problems. The proposed framework is then applied to a set of environmental problems and solutions demonstrating the issues of values, politics, science and technology, policy design, economics, and organizational management.

This multidisciplinary approach owes its origin to Graham Allison's classic work, *The Essence of Decision* (Allison 1971; Allison and Zelikow 1999). Allison posits three models for examining the events of the Cuban missile crisis: the rational actor, organizational process, and governmental politics. He provides different explanations for the events of the crisis, depending on which model he applies to interpret them.[1] In the case of the missile crisis, the "rational actor" model explains the placement of missiles in Cuba as the act of a rational, goal-seeking decision maker. The "governmental politics" model focuses on the political competition among stakeholders for power, thus explaining the placement of missiles and the U.S. response in terms of a competition for political power. Finally, the "organizational process" model highlights the impact of organizational routine and standard operating procedures in constraining the rationality of decision making.

The framework I propose here is not as well developed as the concepts Allison applied; however, it does call for the application of different vantage points when assessing environmental problems, policies, and programs in order to shed light on its different dimensions. In *The Essence of Decision,* Allison provides an analytic method I have always found useful, the image of snapping a lens into place, like the apparatus an optician uses to test improvements in vision based on different prescriptions, so that we may interpret events through the vantage point of that particular lens and, in doing so, bring new facts to light. I borrow this image from Allison and, in a preliminary fashion, apply it to a set of environmental issues.

One purpose of this framework is to counter a deep analytic bias in the way we understand environmental problems. Economists frequently misunderstand the issues of environmental science, ecology, and technology; engineers often ignore the political factors affecting environmental policy; and too many of us simply forget about matters of ethics and values. Paying lip-service to the notion that environmental problems are inherently interdisciplinary does little to amend the tendency to assume that one's own discipline is central. When analyzing an environmental issue, ignoring other fields is an obstacle to better solutions.

The strength of the framework proposed here is that it can be used to understand the causes of environmental problems, the way our society's systemic and institutional policy agenda define them, as well as their evolution over time. Each dimension of the framework illuminates a different aspect of the environmental problem, and, as will be demonstrated through the case studies discussed in later sections, the nature of each problem is weighted more toward certain dimensions than others.

Values

Environmental ethics is the most important of the five dimensions examined here. Ideas about our relationship to the ecological environment derive from our concept of property and a definition of nature as a resource to be used for human material well-being. The domination or taming of the environment has long been a theme in the development of Western politics, economics, society, and religion. It is, in fact, central to what we call "civilization," a term defined as human mastery over other species, and the development of surplus wealth and leisure time needed for thought, reflection, and the transmission of learning. To the extent that we are successful generating the surplus wealth required for civilization, the natural environment is seen as something available for our use, a set of resources to be consumed.

We are more dependent on natural systems than we once thought. We now know that we cannot supplant resources and still maintain a high-quality existence as that notion is currently defined. We need ecological systems. Our technology is not sophisticated enough to do without them. The pragmatic argument is compelling, but it is not the only line of reasoning. For instance, some environmental philosophers suggest that our very arrogance may be at the heart of our environmental problems, that to address these problems we must redefine our relationship with the environment and no longer view other species as resources (Leopold 1949). While this may be true, the planet's more than six billion people will probably not seriously contemplate a return to nature. Other values we hope to achieve such as equity, justice, family, and education also preclude a radical redefinition of our relationship to the biosphere.

Given the current worldwide disparity in wealth, it is difficult to halt economic development and its associated environmental impacts. Instead, some analysts predict that economic development will result in demographic transitions that reduce population growth and increase the public's stake in protecting the environment (Cohen 1995, 47). The idea is that as economic development grows, there will be a decrease in the demand and supply of labor and an increase in the demand and supply of capital. Thus, whereas in developing nations, children (who represent added labor capacity) are perceived as essential for economic survival, in developed nations they are seen as "decorative," an economic liability. As a result, there is less economic incentive to have children in developed nations. According to this theory, only economic development can bring population stability to the planet (Ophuls and Boyan 1992, 46). The language of economic development in recent years incorporated the notion of sustainability, or, put another way, development with sensitivity to environmental impacts. The hope of development advocates is that a fully developed world with low population growth will be less detrimental to environmental quality than the partially developed world we now live in.

The desire for economic development is an expression of values. A good life, as we understand it today, includes a high level of resource consumption. It is unrealistic to assume that this concept will change. Although the Western pattern of consumption may be abhorrent to many in principle, its seductiveness and appeal is a demonstrated fact of modern life. What, then, is the goal of environmental politics and policy? I would argue that it is one that has evolved over the more than thirty-five years since the U.S. Environmental Protection Agency (EPA) was established to deal with the problems of degradation of the natural environment.

Environmentalism in the United States has roots in late-nineteenth-century anti-urbanism, transcendentalism, and the desire to preserve the productivity of the land for future generations (Rubin 2000, 159). Concern for the environment at first was an aesthetic issue and a lifestyle consideration. It reflected a preference for the virtues of an agrarian or rural way of life. Some viewed cities as corrupt and evil in contrast to wide-open green spaces that could cleanse the soul and stimulate virtuous living. When the EPA was created in 1970, it was primarily an agency to combat air and water pollution. Nearly the entire staff in the newly created agency came from air- and water-pollution control units within the Department of Health, Education, and Welfare (HEW). Dirty air and water were regarded vaguely as unhealthy but decidedly as unsightly. As EPA's mission expanded in the 1970s, other issues became important such as managing solid waste that resulted from urban environmental problems. With the passage of the toxic waste cleanup Superfund program in 1980, the environmental issue was defined as a public health issue. Pollution was not only ugly, but it could also make you sick. This human health orientation continued in the 1980s. In the early 1990s the focus shifted to international environmental problems, especially global climate change. As holes in the ozone and global warming were discovered, the definition of the environmental problem expanded to include a concern for the viability of the planet itself.

Despite the changing definitions of the agency's mission, the concern has always been the same: protection of human well-being. We protect the environment so as not to kill the goose that lays the golden eggs. Our taste for golden eggs, namely, economic consumption, continues to grow. The environmental ethic under which we operate requires us to maintain the biosphere for our descendants, not because we care about them but because environmental deterioration reduces our ability to consume the things we desire—wholesome foods, fresh air, clean water, and coastal cities that are not submerged because of global warming.

Whereas some scholars have argued that resolving the environmental problem requires a change in the dominant social and political paradigms, a fundamental shift in how we view politics, the environment, and one another, others contend that a dramatic shift in paradigms is neither necessary nor feasible (Milbrath 1984, 81). Instead, environmental policy should continue to focus on developing less destructive methods for fulfilling the current consumer ethic. Our approach to environmental issues today results in altered consumption patterns, not a reduction in consumption. For example, surfing on the Internet instead of cruising around in a gas-guzzling auto does not mean a reduction in economic consumption and certainly not a reduction in the nation's waste stream. As I will discuss in further detail later, total production of solid waste in the United States has grown from 2.7 pounds per person per day in 1960 to 4.5 pounds per person per day in 2000 (U.S. EPA 1999, 2000). But during that same period recycling grew from 5.6 million tons per year, or less than 10 percent of total wastes, to approximately 70 million tons per year, or 30 percent of the waste stream in 2000. These figures clearly demonstrate that although the patterns of consumption changed, consumption continued to increase.

The environmental ethic that has had the greatest impact in the last three decades, at least in the United States and other Western countries, has been a form of enlightened self-interest. In a value system such as this, environmental protection is not traded off against the value of economic consumption. Although it is actually another form of consumption and does not signify a break in the culture of consumption, the result has generally been a greater popular awareness of environmental issues.

Applying the Values Dimension of the Framework

The various dimensions of the framework may be investigated in any order to broaden our understanding of environmental issues, but I choose to begin with values because, to me, they are fundamental. In applying this dimension, a number of questions need to be examined. Although some may have no answer, all are of use.

- Does the issue stem from a behavior fundamental to our lifestyle?
- Does the problem, or proposed solution, raise issues of right and wrong?
- Does the problem, or proposed solution, require a trade-off between ecological well-being and human well-being?
- Does the process that created the problem, or the proposed solution to the problem, conflict with ethical or religious precepts?
- Does the problem or its proposed solution raise fundamental issues of conflicting values?
- Can the issue be addressed, and progress still made, without confronting the basic value conflict?

Addressing these questions will illuminate the value dimension of the issue, allowing us to place this aspect of the problem in perspective. While all policy issues have a value dimension, the analytic task is to determine how fundamental the value is, how important, and whether it conflicts with other closely held values. Applying the framework helps us to see the environmental problem as a policy issue. Viewing the issue through the values lens can provide insight into the potential intensity of the issue's political conflict and its saliency as a political issue.

Environmental Politics

The environment as a political issue has not resulted in a major shift in the dominant social and political paradigm, but it has added significant considerations to the process of formulating policy in the United States. The environmental issue has made significant demands on our political processes and institutions. Americans have called for political processes that develop a consensus about the definition of environmental quality and make decisions about methods for achieving environmental goals. In the past thirty years this political process has facilitated a high degree of social learning in the United States. This learning process will continue because new information on human-induced change and ecological conditions is continually becoming available. This information will need to be summarized, disseminated, and understood by decision makers and the broader public in order for policy to adapt to the new information and conditions. Environmental politics is closely connected to economic development and worldwide income distribution. Environmental policy is about individual and collective patterns of resource deployment, consumption, and degradation. Simply put, we must learn enough about the biosphere to ensure that in our use of it we do not irreversibly degrade or destroy it. Once we know the types of behaviors required to sustain the environment, we must organize ourselves to perform those behaviors.

This learning process creates winners and losers. The assignment and distribution of benefits and costs triggers political conflict that both impedes and distorts social learning. People and interest groups sometimes present environmental information that is partial or misleading to serve their own particular interests (Sabatier and Jenkins-Smith 1993). Consequently, environmental policy never appears as a seamless progression from scientific discovery to implemented public policy. Rather, it resembles a meandering series of disjointed incremental steps, looking very much like the type of policy making described by David Braybrooke and Charles Lindblom (1963) in *A Strategy of Decision*. The decision-making strategy they described is "remedial," "serial," and "exploratory." Policy makers move away from problems rather than toward solutions. Braybrooke and Lindblom observed that, "analysis and evaluation are socially fragmented, that is, they take place at a very large number of points in a society. Analysis of any given problem area, and of possible policies for solving the problem, are often conducted in a large number of centers" (104).

Many environmental scientists and advocates lament the messiness of this type of policy process and believe it is inadequate to the task of addressing long-term, interconnected, large-scale problems such as protecting the environment. In this view, partial answers cannot address the root causes of environmental problems. Many environmental advocates criticize incremental environmental policy but are unable to suggest truly viable alternatives. It is not likely that people in developed nations will slow down the input of information, reduce consumption, and return to the land. It is unrealistic to eliminate pluralistic, interest-dominated politics. Certainly even a benign environmental totalitarianism is not a viable alternative form of politics. Mass participatory democracy seems equally unlikely, and might not even protect the environment. Meeting the short-term needs of the mass public could pollute the planet, and as orderly as totalitarianism looks on the surface, resistance to authoritarianism is a virtual certainty. In other words, we are stuck with the messy, partial, and incremental politics characteristic of Western democracy. We will need to work within the current political framework if we are going to protect the environment and promote sustainable development.

Even if we were able to achieve a perfect understanding of the environment and the effect of human interaction upon it, our social and political processes cannot absorb and act on the volume and complexity of that information. The exception would be a genuine crisis. Normal politics and incremental policy making can be suspended for a time during a crisis. In the United States during World War II, for example, the government spent nearly 46 percent of the Gross National Product (GNP) and certain civil liberties were suspended (U.S. Department of Commerce, Bureau of Economic Analysis 1994). In the weeks after the World Trade Center was destroyed, partisan politics was replaced by an unusual degree of national unity and patriotism. However, wartime mobilization and crisis management politics cannot be sustained indefinitely. Eventually normal politics resume. The difficulty with the environmental problem is that once we reach a crisis point, it may be too late to solve the problem.

The answer, to the extent that there is one, is to organize politics to accelerate the learning and decision-making process. We need a more rapid incremental political process. Environmental issues must be raised and discussed through the electronic media, public education programs, and active efforts to elicit citizen participation in policy making. Worldwide environmental education has grown exponentially since 1970 (U.S. EPA 1996, 3). In developed countries young people are raised to understand facts about the biosphere that were unknown when today's Baby Boomers were growing up. On the other hand, fear of environmental damage has resulted in the reflexive "not in my backyard" (NIMBY) syndrome that sometimes produces greater total environmental impacts in order to avoid lesser effects on a more powerful or better organized local constituency.

In the United States one result of increased levels of environmental concern and literacy has been a series of successful efforts to protect the environment. As the EPA data indicate in Figure 2.1, pollution in the United States has decreased dramatically as population and GDP growth have continued. The irrational, nonanalytic decision-making and political process in the United States has brought about a successful reduction in key pollutants. This does not mean that the environmental problem is solved or has gone away but rather that we are "moving away from the problem." In the area of solid waste, per capita production of solid waste grew from 2.7 pounds per person in 1960 to a peak of 4.6 pounds per person in 1999 but fell to 4.5 pounds per person in 2000 (U.S. EPA 1999, 2000). I believe, although I cannot say definitively, that the problems of environmental degradation and solid waste management in the United States are slightly less severe in 2006 than in 1994. We recycle more, and continue to reduce releases of key pollutants into the environment. How did we make this progress? What type of political process did the United States engage in? The environment, over time, achieved status on the political agenda. The definition of environmental politics has changed, but it

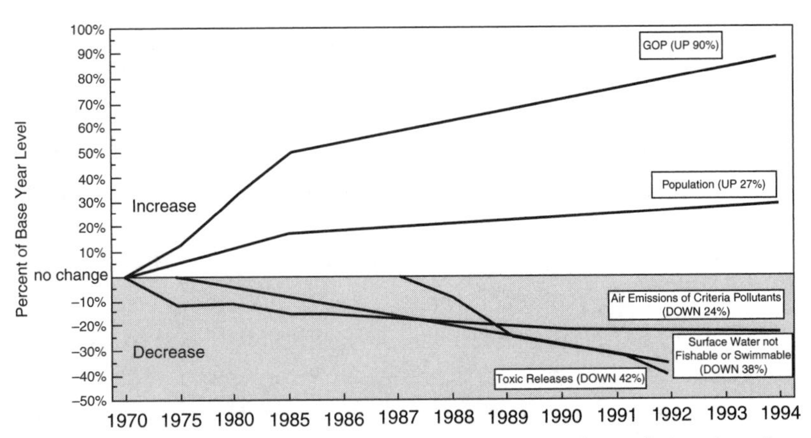

Figure 2.1 Environmental Improvements / Economic and Population Growth

has resisted a number of concerted attacks on its legitimacy, and now it appears to be a permanent fixture on the political agenda.

To understand environmental policy one must understand the agenda-setting process. Why are some issues that reach the agenda acted on whereas others, through a process that has been termed non–decision making, are ignored (Bachrach and Baratz 1970, 44). Issues can also be denied agenda status altogether by denying them legitimacy. Typically, powerful interests in a society define an issue as illegitimate not by responding to it substantively but by using their control of the agenda to ensure it is never heard. For many years issues such as race and gender bias simply could not be heard. In their book, *Participation in American Politics,* Roger Cobb and Charles Elder (1983) took the issue of the policy agenda even further and divided that agenda into systemic and institutional dimensions. An issue that can be discussed obtains status on the systemic agenda as a legitimate issue. It then must "travel" to the institutional agenda where it is seen as a legitimate object of government policy making. Non–decision making is most effective and least visible in keeping items off the systemic agenda, but it also can be used more overtly to keep issues off the institutional agenda (Cobb and Elder 1983).

The environment reached the U.S. national arena in several stages. Rachel Carson's *Silent Spring* (1962) and Barry Commoner's *The Closing Circle* (1971) popularized the concept of a global ecosphere threatened by human economic activity such as the application of pesticides and nuclear testing. After the 1968 presidential election, the environment began to enter the national political agenda in Washington.

During that campaign Senator Edmund Muskie of Maine was so impressive as a vice presidential candidate that he was immediately considered a front-runner for the Democratic presidential nomination in 1972. Muskie's major issue was protecting the environment, and in 1969 and 1970 he pushed for the enactment of an air pollution control act that would set national standards for ambient air quality (Jones 1975, 175–182). Although industry initially opposed the act, President Nixon came to support a national air quality bill to counter Muskie's growing political strength. Nixon and others also saw the environment as a safer, less contentious issue than the war in Vietnam. Some political analysts at the time viewed Nixon's support of the environment as a way to distract people from the Vietnam War, but, even so, it was an important step on the road to formalizing environmental policy and one that was introduced through the American political machine.

The U.S. EPA was established during the Nixon administration by executive order, not through a congressional act. This, along with the enactment of the 1969 National Environmental Policy Act (NEPA) and the 1970 Clean Air Act (CAA), provided President Nixon with an environmental record to counter Senator Muskie's in the 1972 campaign. In the end, Muskie failed as a presidential candidate, and with a weak challenge from George McGovern in 1972, Nixon felt confident enough to veto the Federal Water Pollution Control Act, which was subsequently enacted *over* his veto. Despite Nixon's action on the water bill, in the 1970s the environment was typically seen as a foolproof, popular political issue. With the exception of the early years of Ronald Reagan's presidency, environmental protection has continued to be seen as a straightforward issue for politicians. Nowadays no politico can afford to be perceived as anti-environmental.

In the late 1970s and early 1980s conservative Republicans from the western part of the United States developed an anti-environmental ideology based around the issue of property rights (Layzer 2002, 242). In exchange for their support of Ronald Reagan's presidential campaign, they were given the Department of Interior (James Watt) and EPA (Ann Gorsuch, later Burford) to manage. To Reagan's White House team these were relatively unimportant ministries, so they paid little attention to the selection of the secretary of the interior or the administrator of the EPA. The appointments were used to repay political debts. Moreover, Reagan's senior advisers assumed that popular support for the environment was diminishing. No longer was the environment on the "top-ten list" of issues cited by the American public as "important" in public opinion polling. Environmental interest groups recognized the threat to the gains of the 1970s and organized a campaign of opposition that culminated in the resignation of Ann Burford and her replacement by William Ruckleshaus, the first administrator of the EPA (Kraft and Vig 1984, 3).

It turned out that the Republican political leadership had misread the polling data. The environment had become a less pressing issue among voters because the average American thought that reasonable progress was being made to clean up pollution, not because they had lost interest. When it became clear that the leaders of the EPA and the Department of the Interior were attacking some of the programs that had brought those gains, the rating of the environment's importance in national polls rebounded to levels as high or higher than in the early 1970s (Mitchell 1984, 56). Recognizing that they had landed on the wrong side of the issue, and with the 1984 presidential election approaching, the Reagan administration moderated its views, forced out some of the most visible right-wing environmental leaders, and allowed gradual environmental progress to resume.

In the first decade of the twenty-first century President George W. Bush also struggled with his image on environmental issues. In early 2002 the American public was not sure where President Bush stood on environmental issues. He was not yet defined in the public mind as anti-environment. Christie Todd Whitman, Bush's first EPA administrator and the former governor of New Jersey, was a moderate with a good environmental record. Her appointment to the EPA indicated that the lessons of the mid-1980s had not been lost on President Bush and his political team. His second administrator, former Utah governor Michael O. Leavitt, was also an environmental moderate. His third administrator, Stephen Johnson, was an environmental scientist and a twenty-year veteran of the EPA. By 2003, however, Bush began to slip into an anti-environmental position.

Indeed, by the end of 2003, George W. Bush had become the most anti-environmental president since the creation of the EPA. At the start of his second term, this pattern continued, with occasional efforts to paint the president as an environmentalist. Although his administration came into office determined to avoid the mistakes of the Reagan years regarding environmental policy and politics, it was unable to do so. Instead, the administration began to launch a series of subtle attacks on the rules regulating environmental protection, from the "Clear Skies" legislation that would slow the national air cleanup effort to the "Healthy Forests" initiative that would damage the nation's wildlife and forest ecosystems. Congressman Tom Allen of Maine called Clear Skies "a classic case of chutzpah, a triumph of marketing over substance" (2004). Sierra Club president Carl Pope commented that the Healthy Forests Restoration Act "will certainly succeed in propping up the timber industry" (Jalonick 2003). The public relations terminology used to describe these bills together with revelations in 2003 of White House efforts to tone down the EPA's warnings about air pollution resulting from the destruction of the World Trade Center are examples of the administration's lack of fervor on environmental issues.

In President Bush's first term, not only did we witness an effort to reduce the scope of environmental law, we also saw the attempt to diminish the enforcement of existing laws. On November 5, 2003, the EPA dropped investigations into fifty power plants for past violations of the Clean Air Act. A scaling back of that act that at the time was predicted to reduce investment in cleanup equipment by between $10 and $20 billion. In response, New York State Attorney General Eliot Spitzer, along with a number of his colleagues in other northeastern states, decided to continue legal action, without the federal government, to force these utilities to clean their emissions (Barcott, 2004, 38). Although most Americans favored the Bush administration's environmental stance in 2001 and 2002, support for its environmental leadership fell to 44 percent in April 2003 and continued to decline to 41 percent in 2004 and to 39 percent in 2005 (Gallup Organization 2005b).

The environment's political power tends to lie dormant as long as the public perceives that leaders are serious about promoting pro-environmental policies. But once a leader is no longer credible on the environment, the issue gains in urgency and importance, and can rise in national polls. The controversy over the nominations of both Utah governor Michael Leavitt in 2003 and Stephen Johnson in 2005 as the EPA administrator reflected the president's reduced credibility on the environment. Even though Governor Leavitt was considered an environmental moderate and Johnson was a well-regarded environmental professional, Democrats and environmentalists delayed their appointments as a way to focus attention on the administration's weak environmental record (Stout 2005).

Environmental issues do not typically decide elections, because most people assume that just as all elected leaders promote security and safety, they also promote environmental protection. The need to breathe clean air and drink safe water is not politically controversial. Political analysts often confuse the lack of passion behind the issue with a lack of public interest or concern. That is a mistake. Even on tough trade-off questions the public consistently supports environmental protection. In Gallup polls since the 1980s to 2001, the public, by a 2 to 1 margin, has favored policies promoting environmental protection even at the expense of economic growth. Through 2000, by a 3 to 1 margin, the public believes that the environmental movement in the United States has done more good than harm (Gallup Organization 2005c). Public support for environmental protection runs deep in this country, cutting across every demographic category (Gallup Organization 2005a) .

Nonetheless, as long as the public sees that good-faith efforts are being made to protect the environment, the issue does not generate much political heat beyond the environmental community. Similarly police protection and education are also nonissues until the public perceives they are threatened. However, when the public believes that a key value such as education, police, or environmental protection is threatened, the political reaction is rapid and sweeping.

An oddity of President Bush's approach to environmental issues is that it reflects an antiquated understanding of the environment. This notion is at the heart of the Gallup poll question trading off environmental protection against economic growth. Although there is a connection between environmental quality and economic growth, it is not a trade-off relationship. While some corporations might find that pollution control costs them short-term profits, in the long run investments in environmental protection tend to pay off. We are learning that economic growth *depends* on environmental quality: the return on investments in environmental quality pays off. Sound and sustainable economic development requires the maintenance or restoration of environmental quality. A study published by the Office of Management and Budget (OMB 2003) estimated that from 1993 to 2003 we spent $23 to $26 billion to clean up our air, resulting in net benefits valued at between $120 and $193 billion for that investment. Investments in sewage treatment have increased the value of waterfront property all over the United States. The funds given to property owners to reduce pollution near New York City's reservoirs have saved the city $6 billion that it would have had to spend on a water filtration plant (Office of New York State Attorney General Eliot Spitzer 2003).

Popular support for protecting environmental quality is the basis for the issue's political strength. The importance of environmental protection has been the subject of massive educational and propaganda efforts by scientists, advocates, the media, and professional educators. Although respect for property rights remains strong in the American political culture, especially in the western states, support for environmental protection frequently dominates concern for property rights. Indeed, one source of political strength for environmental protection is the perception that pollution can diminish the value of private property. In this respect, high environmental quality is a form of property or wealth. This did not happen overnight and is arguably a result of the high level of economic wealth in the United States. Our wealth permits consumption of beach homes, country homes, suburban living, and vacations in national parks and rural areas—"goods" that environmental degradation would diminish. Maintaining that level of consumption requires the maintenance of environmental quality.

Despite support for curbing the pollution of others, not all consumption patterns in the United States provide behavioral evidence of support for protecting the environment. The increased average weight of the American automobile is an example of behavior that does not promote environmental protection. Nevertheless, the popular consensus for protecting the environment in the United States is strong across regional, racial, and socioeconomic categories (Mohai and Bryant 1998, 475). Clearly this type of consensus facilitates social learning about the environment, eventually leading to effective policy making.

An important dimension of the environmental problem is its status as a political issue. This involves its presence on the political agenda as a problem appropriate to collective societal action.

Applying the Political Dimension of the Framework

As the preceding discussion indicates, an issue only becomes one of public policy when it is part of a political process. How the issue is defined, how it enters the political agenda, and the views of its key stakeholders determine how that issue is defined. Once again, in applying this dimension, a number of questions need to be posed, many of them unanswerable but still essential:

- What is the status of the issue on the policy agenda?
- What is the issue's degree of legitimacy?
- What role has the issue played in electoral politics?
- Who are the political stakeholders involved in the issue, and what is the nature and style of their participation? How important is the issue to these stakeholders?
- Who are the potential winners and losers in the political competition around the issue? Who have been the winners and losers to date, and how is political victory and loss defined in this issue area?

- To what degree are stakeholders willing to discuss and compromise on the issue?
- Does the issue act independently of other political issues, or does it cluster with other key issues?[2]
- In the U.S. context, which level of government is considered primarily responsible for addressing the issue (state, federal or local)?
- What is the level of controversy and consensus around the issue? What are the areas of agreement and disagreement?
- How does scientific certainty or uncertainty related to the definition of the problem or its potential solution influence the politics of the issue?

The political definition of the issue is critical in shaping how the issue is framed and ultimately addressed. For example, a nuclear power plant may be seen as a source of vital electric energy, as a terrorist target, as a source of potential contamination from a meltdown, or as a source of a waste product difficult to discard. It can also be seen as a combination of some or all those factors. The early part of a political controversy is often a struggle over the definition of the issue. That definition shapes the way the problem is perceived and sets the boundaries for the problem's solution. The definition will persist, often in the face of new scientific information, a different context, and changed behavior. Understanding an environmental policy issue requires a careful and sophisticated analysis of the issue's political dimension.

Science, Technology, and the Environment

Much of the progress we have made in protecting the environment has been the result of the development and implementation of technological fixes to environmental problems. We reduce air pollution by utilizing newly developed environmental controls such as catalytic converters on autos and scrubbers on electrical power plants; we treat sewage before dumping it into waterways. In short, we use science to fix the mess that science helped develop. The question is, can we solve problems as quickly as we create them? The answer is no. The question then becomes, can we fix the most pressing problems fast enough to maintain a habitable environment? Here the answer is more complicated. What do we consider a habitable environment? Are the shanty towns and slums of developing countries livable? When is an environment so dangerous that we consider it unacceptable? If we create a technology that causes disease in some percentage of people, but we develop a cure for that disease, is our tolerance for a lethal environment expanded?

To some degree, then, the environment is a problem of science and technology. We invent new products and put them to use before we project their effect on human health, the biosphere, and the local ecosystem. Until now, technologically based economic development has raised living standards and increased population around the globe. The benefits are unevenly distributed, but the results are undeniable. Can science and technology keep up? When science cannot develop remedies to dangers caused by new technologies, can we slow down the introduction of additional technologies until we figure out how to use them safely?

Experience over the past thirty years provides evidence on both sides of this issue. In the United States the problem of nuclear waste and reactor safety has limited the use of nuclear power to generate electricity. A number of toxic substances, such as the insecticide commonly known as DDT, have been banned here. Chlorofluorocarbons, the refrigerants that cause damage to the ozone layer of the atmosphere, are gradually being replaced. Yet some developed nations, including France, continue to rely on nuclear power for electricity, and many developing nations still use DDT as a pesticide. My view is that if the harm is easily proven and a clear technical fix is in place, with strong government intervention technological solutions can be implemented. If the price of the substitute is too high or the technology not fully developed, eliminating a dangerous technology is more difficult. For example, if fossil fuels were not relatively plentiful at the moment, the risk of nuclear power would be given less attention than its benefits.

Problems caused by the impact of technological innovation on the environment are not easy to measure. The problems may take a long time to develop, and sometimes it is hard to establish a causal relationship between an environmental problem and the introduction of a specific technology. Even more difficult to assess are problems caused by the interactions of one or more technologies in varied ecological settings. Environmental impacts are unavoidable by-products of the strength and power of the scientific method.

The scientific method is based on the concept of the controlled experiment. The researcher first isolates the variables, only subjecting certain ones to a particular test. The goal is to identify and understand causal relationships, "all things held equal." But in the natural ecological environment, nothing is held equal. The interactions and relationships that take place can best be understood at the system level where controlled experiments are rare. Whereas experiments in the traditional scientific method are reductionist, that is, an attempt is made to reduce the test to a simple causal relationship, ecological systems are holistic and interconnected, and cannot be understood by reducing reality into simple relational terms. Such understanding requires the use of models that estimate the interaction effects and account for the multiple feedback loops that characterize living systems. Technologies developed with reductionist methodologies must then be analyzed with environmental impact studies that are based on a more appropriate, system-level conceptual framework and orientation.

In a controlled experiment, the whole test is designed to demonstrate or rule out an effect. When an effect is discovered and verified, it then becomes a fact. When a model is built, we leave the world of scientific certainty and enter one of probability statements and other unknowns. It is difficult for an environmental scientist who suspects a harmful effect to compete with the power and the certainty of a technology's proven benefit. When an environmental modeler thinks he or she may have uncovered a destructive effect, it is initially expressed as a probability statement. In fact, the most persuasive scientific evidence of environmental damage must rely on models to develop hypotheses for relationships that are then tested in reductionist, controlled experiments. Only at that point can the factual basis for environmental damage be conclusively established.

Improvements in environmental measurement technology and in environmental modeling provide some hope that we are still learning and will do a better job of detecting, understanding, and ending practices that harm the environment. The information provided about environmental damage, however, is not always factored into decision making. To the degree that the public is educated about environmental threats, public opinion can become a powerful force behind active environmental protection policy. But many of the threats to the environment are long term, difficult to prove, and hard to explain.

Applying the Science and Technology Dimension of the Framework

Some environmental problems, such as the Indian Ocean tsunami in late December 2004 or Hurricane Katrina in 2005, are the result of natural phenomena but most are human made. As we seek to understand an environmental issue, an important dimension to consider is the level of scientific knowledge and certainty associated with the problem and its potential solution. As we seek to understand this dimension of the problem by applying this element of the framework, we attempt to address the following questions:

- Is there scientific certainty about the causes and effects of the problem?
- What are the principle areas of uncertainty, and what is the effect of that degree of uncertainty on decision makers?
- Are there cost-effective substitutes for the technologies causing harm? What are the prospects for developing such technologies?
- Does the technological cause of the problem need to be halted in order to address the problem?
- Are "off-the-shelf," proven technologies available to mitigate the impact of the environmental problem? What are the prospects for developing such technologies?
- Are the control or mitigation technologies widely available, and do we have experience with their management?
- What is the monetary cost of research to address this issue, and are these funds likely to become available?

Some environmental issues are scientifically complex, and others are simple. Some lend themselves to relatively low-cost technological fixes, whereas others are expensive to address. The relative scientific certainty, complexity, and potential cost of control or remediation technology influences the political definition of the issue.

Climate change, discussed later in this book, provides an excellent illustration of the role scientific uncertainty plays in defining a policy issue. In the first term of the George W. Bush administration, inaction on climate was based on a perceived lack of scientific certainty regarding the causes of global warming. As that uncertainty was reduced and scientific consensus emerged, pressure to address the issue grew. In 2005, the first year of Bush's second term, this pressure continued to increase. The level of scientific uncertainty can therefore influence the type of policy design that is appropriate to address an environmental issue (Layzer 2002, 230). Nearly all environmental issues go though an early "problem definition" phase where scientific research is funded to increase our understanding of the problem, thus delaying action. Water pollution and air pollution policy in the 1950s and 1960s focused on increasing our understanding of the science of the problem and potential solutions. Often when environmental programs are established, control technology is fairly primitive. Policy makers hope that the need to comply with new environmental standards will force the development of new technology. Often new environmental rules have had that very effect.

Environmental Policy Design and Economic Factors as an Influence on Detrimental Corporate and Private Behaviors[3]

Economic forces are the major cause of environmental degradation and are the primary means of environmental preservation, cleanup, and pollution prevention (Schneider and Ingram 1997, 99; Stroup and Shaw 1989). The environment as a public policy issue should be conceptualized as a form of government regulation of corporate and individual behavior. This section deals primarily with the design of policies that regulate corporate behavior, since that has been the main target of environmental policy and regulation thus far. Toward the end of this section, the regulation of individual behavior and the problem of social learning are discussed.

Certain elements of the framework helped to explain environmental problems and solutions, but here we focus on understanding environmental policies (solutions). The policy approach that is taken relates, of course, to the definition of the problem. The policy approach can also influence the evolution of the problem. If progress is made, the problem may come to be seen as routine and less urgent. This section provides a catalogue of the variety of policy designs that have been used to solve or address environmental problems.

In the past three decades we have heard a good deal of political, popular, and academic discussion on the concept of regulation (Litan and Nordhaus 1983; Pedersen 1991). Regulation is criticized as harming the economy by stifling entrepreneurial initiative, discouraging technological advances, and not being sufficiently cost-effective (Bardach and Kagan 1982; Johnston 1991). Economists criticize lawyers for being overly formalistic and for not understanding how firms behave. Policy makers disparage economists for making proposals that lack political feasibility.

Defining Regulation

Kenneth Meir (1985, 1) defined regulation as "any attempt by the government to control the behavior of citizens, corporations, or sub-governments." Regulation is a set of rules or directives intended to induce specific behaviors in target populations. Modifying Meir's definition slightly, substitute the word "influence" for "control." Regulated behaviors represent tendencies and carefully augmented actions rather than goal seeking, rationally controlled behaviors. Control is simply too strong a term. Organizations for the most part do not truly control their own actions; instead, these actions are the result of a variety of internal exchange relationships and influences evidenced by explicit and implicit bargains and the deployment of potential and actual incentives. Again, this regulated behavior is merely a tendency to incremental actions rather than goal seeking, rationally controlled behavior.

The goal of regulation is to influence individual or organizational behavior. To provide a graphic example, consider the case of automobiles converging on a corner traffic light. One hopes that the behavior of each driver is influenced by the color of the traffic light. The signal is relatively clear, but when the light turns amber, drivers are faced with the need to make a rapid decision whether to slow down, speed up, or stop. Several factors affect each driver's decision:

- Is the signal working?
- Does the driver see and understand the signal?
- Is the driver willing to adhere to the signal?
- Is the car mechanically capable of stopping or accelerating?

Are the regulated parties, in this case the drivers, capable of changing behavior in the desired direction, and are they willing to do so? The goal of regulation is to influence the variables that enter into a regulated party's calculus of the costs and benefits of compliance. What are the incentives to stopping at a red light?

- The possibility of a collision with a fully loaded trailer truck
- A traffic ticket from a highway patrol officer for running the light
- Belief in the rule of law
- Pre-patterned behavior of braking for a red light

And what are the disincentives to stop?

- A severely ill child in the back seat, and an urgent need to get to the hospital
- One is in a hurry, and no traffic is visible.

The goal of traffic regulation is to reinforce the incentives to comply so that they outweigh the potential motivation to pass the red light. Similarly, the goal of regulation is to influence the perceptions and behaviors of regulated parties. Each regulatory program, therefore, must be based on a strategy that seeks to understand the motivations of regulated parties in order to influence their behavior.

Policy Design: How to Develop and Implement a Regulatory Strategy

Strategic regulatory planning is an effort by government to develop a comprehensive strategy for influencing behavior (Cohen and Kamieniecki 1991, 12–13). There are two components to this plan. The first is the formal regulation itself; the second is the manner in which the regulatory plan is implemented. Extra-regulatory elements that can be manipulated to encourage compliance include funding, technical assistance, exhortation, and publicity. Since willingness and capacity to comply with regulation can vary widely within a given regulated community, having an array of regulatory mechanisms available is critical. It is also important to approach the task of influencing behavior without ideology or preconceptions.

One might argue that it is administratively or legally infeasible to target regulation for maximum influence on specific regulated parties.

The administrative argument is easy to counter. First, regulations are now individually tailored through the permit process (Rabe 2000, 36–37). Second, it is possible to deal with groups of regulated parties and tailor approaches to classes of regulatory situations rather than to individual organizations. Finally, an approach focused on changing the behavior of regulated parties will tend to be less process-oriented and thereby less administratively complex. It will also utilize strategic alliances between different parties who share a similar interest in the successful implementation of the regulatory program.

The issue of legal feasibility is the argument that the law cannot be adjusted to account for an organization's willingness and capacity to conform to the law's requirements. Regulatory enforcement through the courts, one should note, typically results in bargains that take into account an organization's capacity and apparent willingness to move toward compliance. That the application of environmental rules involves these negotiations needs to be acknowledged. The notion that the law is applied without consideration of feasibility is simply untrue. In fact, Cass Sunstein (1990, 416) argues that when regulators are compelled to implement rules that do not allow them to consider issues of feasibility, they frequently fail to act. A more typical response than inaction is deal negotiation. Frequently this involves a compliance schedule and other government concessions.

A strategic approach to regulation would acknowledge the reality of the bargaining process up-front and develop compliance strategies with input from the regulated community. Under these circumstances, enforcement and the threat of enforcement is reserved for recalcitrant organizations that willfully violate agreements, engage in deception, or are otherwise averse to changing their practices.

The Tools of Strategic Regulation

"Command and control" describes a process where government commands a regulated party to act in a certain way and then uses the legal system to control behaviors that are not in compliance with the rules. The traditional notion of this approach is a simplistic view of regulation, for regulation involves all government policies and programs deployed to influence the behavior of regulated parties. An updated definition of regulation includes command-and-control regulation along with the use of market mechanisms and a wide variety of other techniques of influence.

One need not choose between command-and-control and market mechanisms. Neither one is necessarily better than the other (Ackerman and Stewart 1988, 171). Each target of regulation must be assessed to determine the mix of incentives and disincentives that will result in the desired change in behavior (Rosenbaum 2005, 167). Various techniques of influence are available to government regulators:

- Market solutions and economic incentives
- Insurance programs
- Self-regulation
- Taxes and fees
- Education, information disclosure, and the use of the media
- Tracking and reporting formal compliance
- Licensing
- Permitting
- Setting standards
- Grants, training, and compliance assistance
- Assessing penalties
- Inspections
- Adjudication

These activities include both coercive and relatively noncoercive actions. Policy design should favor the least coercive methods that obtain the desired results. The following regulatory actions comprise a partial listing of activities typically available to regulators that influence the behavior of regulated parties.

MARKET SOLUTIONS AND ECONOMIC INCENTIVES

Government sells firms and other private parties permits to pollute, specifying an allowable level of pollution. These permit levels can be traded to other firms, creating a market in pollution allowances. This encourages permit holders to reduce their own level of pollution and maximize the cost-effectiveness of pollution control (Hahn and Hester 1989).

INSURANCE PROGRAMS

Government requires private parties to carry insurance in order to clean up unanticipated releases of pollution and to compensate victims of negative environmental impacts. For example, a gas station owner may be required to carry insurance to pay for the cost of cleaning up gasoline leaks and to pay third-party liability claims arising from such leaks.

SELF-REGULATION

Government permits an industry to regulate itself. The use of industry codes and professional ethics are examples of such self-regulation.

TAXES AND FEES

Taxes on activities that pollute can be used to generate revenues to provide financial incentives for activities that do not pollute (Levenson and Gordon 1990). Government charges regulated parties for each unit of pollution or waste created. Alternatively, raw materials that eventually cause pollution are taxed, as in the pre-1995 Superfund's tax on oil and chemical feedstocks.

EDUCATION, INFORMATION DISCLOSURE, AND THE USE OF THE MEDIA

Government informs the public about regulatory violations or dangers, causing a company negative public relations. An example is the warning label requirement on cigarettes. Government may also use the media to educate the public about regulatory requirements and their purposes.

TRACKING AND REPORTING FORMAL COMPLIANCE

Government rules require regulated parties to report on their compliance, which is less expensive than inspections and can serve to begin the process of creating institutional capacity in regulated firms to comply with a rule. Whoever fills out the form must at least pay some attention to the regulation.

LICENSING

Government certifies competent professionals who can assist with compliance. The best example of this method is the regulation or licensing of certified public accountants, who facilitate compliance with tax regulations. In the environmental area it may be possible to certify environmental auditors and other professionals who could help a firm reduce and prevent pollution.

PERMITTING

Government requires firms to obtain a permit in order to pollute legally. A permit can call for gradual reductions in pollution. The absence of a permit can result in a judicial order to close a factory.

SETTING STANDARDS

Setting standards is the traditional command element of command-and-control regulation (Breyer 1982, 96). There are two basic types of standards. The first is the performance standard, which requires that specific goals be accomplished but does not specify how those goals are achieved. A second type of standard specifies a process, technology, or practice that a regulated party must deploy to be in compliance with a rule. This simplifies regulatory compliance, and oversight of compliance, by requiring a specific, easily measurable activity, but it also reduces the discretion a firm has to determine the most cost-effective mode of conformity.

GRANTS, TRAINING, AND COMPLIANCE ASSISTANCE

Many of the targets of regulation are individuals and small businesses that are willing to comply but lack the capability or resources to do so. Sometimes grants, loans, or even loan guarantees can help a small business obtain the capital needed to comply with a regulation. Training and consulting services can also have a large impact, especially in areas where regulation and technologies are new.

ASSESSING PENALTIES

Typically, penalties are fines charged against violators. Penalties are particularly complex disincentives and must be used with great care. A penalty that is too low is simply absorbed into the cost of doing business. One that is too high can result in extensive litigation and steep transaction costs for the agency. It can also lead to illicit avoidance behavior or political opposition to the legitimacy of the regulation and even the regulator. Nevertheless, as the Internal Revenue Service (IRS) has learned by auditing celebrities and as Eliot Spitzer has demonstrated by prosecuting large corporations, a well-targeted penalty with sufficient publicity can result in widespread compliance to an agency's rules.

INSPECTIONS

Visits by regulators to regulated parties to determine compliance are an important part of the traditional command-and-control model.

Inspections provide evidence that regulated parties are following the rules. A more important use of inspections, especially if they are random and unannounced, is to stimulate compliant behavior for fear of an impending inspection. How many people keep careful tax records for fear that one day they will be audited by the IRS?

ADJUDICATION

Formal adjudication is an administrative or judicial trial to determine if a regulated party has violated a rule. The threat of adjudication often promotes compliant behavior.

A Strategic Approach to Regulation

The choice between command-and-control and market-based regulation is a false one. All regulation involves gradual, strategic calculation and bargaining. The command-and-control strategy results in regulations that adjust the law to reality, permits that interpret regulations in the light of practical constraints, and judicial and administrative bargains on how permits should actually be implemented. Donald Elliot, former EPA general counsel, notes:

> It is important to recognize that we don't have to have and we don't have an all or nothing system in which we have either an incentive-based system or a health-based system of command and control regulations. Many of our environmental problems, like many of our other legal problems, involve a complex coming together of different goals and different moral norms. The system cannot simply optimize any single value like controlling the total amount of pollution at the least cost but must be responsive to multiple values. Multiple goals for hybrid systems . . . Thus a combination of health-based standards and market-based incentives may be preferable to either standing alone. (Breger et al. 1991, 479)

A broader framework is needed that provides policy makers with a menu of devices depending on what and who is being regulated. Some substances are so toxic that command and control is necessary. Some regulated parties are so weak that if they are not paid to comply, they will be driven out of business. In other cases a market can be created, and environmental improvement accomplished, through this mechanism (e.g., recycling or air emissions). The economic causes of environmental problems and the economic impact of proposed solutions vary according to the role the polluting business plays in the nation's economy. The approach to policy design should be as varied as the economic forces the policy is seeking to influence. In some cases, market mechanisms can encourage compliance and avoid the legal and administrative costs of direct regulation. When regulations are necessary, government should provide subsidies, training, and consulting services for organizations that do not have the capacity to comply. Occasionally government may decide that the costs of subsidizing regulation are so high, and the benefits of regulation so important, that a business should be allowed to die to protect the environment. These instances should be as infrequent as possible, or political support for environmental protection will ultimately erode.

Policy analysts often lament that environmental goals are sold to the public with fear and inadequate risk assessment, and to politicians for their value as "pork." They argue that the goals of legislation and regulation ought to be based on careful scientific consideration of risks (Landy et al. 1990, 279–283). Similarly economists frequently argue that policy designs should reflect a careful assessment of costs and benefits, and seek to achieve the maximum possible bang for the minimum possible buck. These ideas seem rational and attractive but are not always feasible in the messy, pluralistic, federal, divided political system we operate in. Sometimes cost-benefit analysis is difficult to conduct. One problem is that the distribution of costs and benefits can be unpredictable, and distribution effects are more politically salient than the overall economic effect. Another difficulty is that some costs and benefits cannot be compared without questionable assumptions about the relative weights assigned to specific cost and benefit factors (Layzer 2002, 8).

There are no shortcuts. Each regulatory program must be based on a strategy that seeks to understand the motivations of regulated parties. Whether we decide to employ direct regulation, indirect market mechanisms, or direct subsidies, none of these approaches will work without a profound understanding of the firms being regulated. Developing the administrative capacity in government to make these assessments is far more important than making decisions on which regulatory mechanism is superior. With this knowledge in hand, environmental regulators can develop flexible and dynamic strategies to reduce and prevent pollution in the real world (Cohen, Kamieniecki, and Cahn 2005).

Policy Design that Regulates Individual Behavior and Stimulates Social Learning

To some degree, regulating corporate behavior regulates individual behavior, and if the corporation is large enough the impact can be massive. For example, by regulating the pollution produced by all cars manufactured by a single large automaker, government has the ability to change the individual behaviors of all those who drive cars made by that company. The compliance of a small number of parties representing a large number of individuals eases the administrative costs and challenges of convincing millions of people to behave in new ways.

Not all environmental problems can be addressed through the regulation of corporations, however. Some environmental policies involve reaching individuals, educating them, influencing their values, and changing their behavior. For example, solid waste reduction and recycling both require a change in individual behavior. To recycle, people need to sort their garbage within the household. Even if the technology of garbage sorting advances, public understanding of the importance of recycling is needed to ensure that government continues to sort waste for reuse.

Most important is that individuals learn to value the natural environment. Although it is true that valuing the environment is expressed as part of Western consumer culture, there is no requirement that people continue to consume the "economic good" of environmental quality. Living without nature may sound like science fiction, but that people go camping, visit the beach, and enjoy nature is an expression of learned values, not a form of innate behavior. If we stop valuing environmental quality and no longer pass that value on to our children, the environment will not generate support in the political or economic marketplace. People may decide to experience nature as a virtual rather than physical reality. While a world based totally on technology may appeal to some confirmed urbanites, we do not have the technology or energy required at this time to supplant natural systems with human-made systems. Our survival depends, therefore, on the use of natural systems to generate our sustenance. If we are to survive, the value of environmental protection must be learned at the individual level.

The levers to inspire this social erudition include economic incentives such as price mechanisms that help people learn about and value the environment (Kolstad 2000). Other tools for social learning are educational curricula and the mass media. All have been used and are still needed to reinforce the message that environmental protection is important.

Applying the Policy Design Dimension of the Framework

A key dimension of a public policy issue is the approach taken by the polity to address the issue. What type of policy design is considered feasible is an indication of the seriousness of the issue, its salience and importance. To understand the policy design dimension of the issue, once again a number of questions can be posed. The aim is to understand the rationale for the approach taken and how the approach has evolved. As noted above, in the early days of air pollution policy, the government convened conferences of scientists and policy makers to discuss the nature of the issue. This very soft approach was suddenly and dramatically modified in the federal 1970 Clean Air Act with the development of the first national ambient air-quality standards. This leads to the first policy design question to address: What is the degree of compulsion and coercion included in the policy design?

Other key questions include the following:

- What is the mix of incentives and disincentives used to influence behavior to reduce damaging the environment?
- What are the economic costs and benefits of the policy design?
- Does the policy design reflect strategic thinking, or is it based on political considerations, stakeholder compromises, or a lucky guess (what Jones [1974, 438] referred to as "speculative augmentation")?
- Does the regulated community understand what they are being asked to do, and are they supportive of the approach taken?
- Are the regulated parties willing to comply with the policy as designed, or will they resist by pursuing noncompliant strategies such as legal challenges and pro-forma compliance?
- Are there other stakeholders who are not regulated parties, and do they support the policy design that has been promulgated?

- What resources are available to ensure compliance with the policy design, and are these resources likely to be sufficient?
- What is the general role of the government and of specific governmental levels and units in implementing the policy design?
- What type of progress away from the problem or toward a solution is the policy design likely to generate, and why?

The design of the policy helps analysts and practitioners understand the operational definition of the environmental problem. The proposed or adopted solution tells you what part of the problem is considered important enough to be addressed by policy makers. The operational definition of the *problem* is the one that the *policy design* seeks to address, just as the operational definition of the *policy* is the *program* that the management system actually puts into place. Let us turn to that final dimension of the framework, the issue of management.

Environmental Management

Once the political dust settles and a policy design is adopted, the environment becomes a management issue. For policy to become meaningful in the real world it must be translated from words to deeds. Policy and management are related. Cumbersome, complex policy designs are typically more difficult to implement than simple designs. Jeffrey Pressman and Aaron Wildavsky demonstrated that point in their classic work, *Implementation* (1984). Policy designs that exhort or mandate private action are less certain to be carried out than policy designs that provide concrete incentives or punishments for private actions.

However, even the simplest policy designs can be wrecked through poor management or political interference. For example, take the case of the Federal Water Pollution Control Act of 1972. Section 208 called for "area-wide regional water quality management plans." Section 201 provided a grant program to help local governments build sewage treatment plants. The idea was that "208 plans" would be used to determine the best place to locate sewage treatment plants, and the selection of plant sites would be based on environmental rather than political criteria. Unfortunately President Nixon impounded the funds authorized and allocated by Congress to pay for Section 208 planning grants.

Because of the lobbying efforts of local governments and construction companies, he did not succeed in impounding the funds for building sewage treatment plants. Billions of dollars were spent in the 1970s on plants that were sited without assessing their role in regional water quality management. As a result, some plants were located in inappropriate places. Some communities overbuilt capacity and needed to attract development to help pay the cost of plant operation and maintenance. Other areas found their growth and development stalled because of inadequate waste treatment capacity.

Despite the lack of planning and the mistakes made, the sewage treatment program was a great success. At its peak in 1976, the federal government spent $9 billion annually on grants to local governments to build sewage treatment plants. This number gradually declined to about $2 billion per year in the 1980s. At that time the grant program was replaced by a State Revolving Fund Program that provided low-interest loans to cities and other governments for environmental infrastructure needs including sewage treatment and systems to control nonpoint sources of pollution. The result of the sewage treatment program and regulation of industrial discharges of pollution is obvious and measurable. In 1974 EPA data indicated that only 40 percent of rivers in the United States were safe for swimming and fishing. Today 60 percent of the rivers are safe for these purposes (U.S. EPA 2003a).

How did the U.S. government organize itself to accomplish these results? First, a decentralized federal structure was put in place. The EPA helped stimulate the creation of state-level environmental organizations, and these in turn encouraged the development of local units. Early in the process of regulating industrial polluters, the EPA delegated enforcement implementation to the states. Policy was developed in Washington, but state and local governments carried out the actual monitoring and enforcement. This helped to ensure that national rules were interpreted to accommodate local political reality. Although this may have slowed down initial efforts at pollution control, over the long run states, with occasional prodding from the EPA's regional and headquarters offices, achieved results.

In the case of municipal sewage treatment, the federal government designed specifications for the plants, recommended contractors, reviewed their competence, and also provided significant funding. The strategy of eliciting state and local buy-in through grants worked.

Although the federal share of costs for these sewage facilities was relatively small in the long run, initially few governments could resist the "free" federal funding. Ultimately the operation and maintenance of these facilities was more costly than the initial capital outlay. The growth of the average American homeowner's water bill in the 1980s and 1990s was a direct result of the need to pay these ongoing costs. It is unlikely, or at least less likely, that sufficient political support for sewage treatment would have been generated if people had fully known and understood the total cost of constructing and operating sewage treatment plants (Freeman 1990, 97).

From a management perspective, the strategy to control water pollution had some useful features. First, it focused on the biggest source of pollution—municipal sewage and the pollution by large industrial facilities. Only a few actors needed to change their behavior to implement the program. The decentralized structure and use of private contractors ensured that centralized bureaucratic clearance was not required during the construction of sewage treatment plants. The public works approach had the advantage of visible, concrete (excuse the pun) accomplishments.

For policy words to become policy deeds, goals must be clear and well understood. The necessary tasks must be simple. Joint actions between organizations and even between individuals should be minimized. The technologies required to implement policy should be well developed and available off the shelf. If new technology is needed, not only must it be invented and then de-bugged for practical use, it must also be explained to those expected to require it, install it, maintain it, and monitor its functioning.

One difficulty with the management of environmental policy is that environmental programs tend to take a piece of a larger problem and subdivide it in order to work on a solution. For example, we use sewage treatment plants to reduce the amount of raw sewage we dump into the water; however, the treatment plant creates a sludge that must either be dumped in a landfill or the sea or burned. Solving a problem in one environmental medium can create new problems in other media. Our hope is that, by gaining a measure of control over the process of releasing the pollutant into the environment, we can minimize the damage it might cause. The analysis of management effectiveness and efficiency must move beyond the performance of the organization and its narrow task; it must be broadened to consider management of the entire ecological system being maintained. Measures of these environmental outcomes need to be developed and used to influence management decisions.

Finally, there is the question of goals and the definition of success. The 1972 Federal Water Pollution Control Act set the goal of ending discharges of pollution into navigable waters by 1977. A nice thought, perhaps, but a ridiculous, unachievable goal. The 1980 Superfund toxic waste cleanup program had a similar problem in goal setting. After a decade of hard work and billions of dollars of expenditure, the press reported that fewer than a dozen toxic waste dumps had been "cleaned up." The press did not report, however, that more than three thousand threat removal and emergency response actions—which are explained in greater detail in chapter 4—had taken place, and millions of people had been moved out of pathways of potential exposure to toxic chemicals (U.S. EPA 1992). Unfortunately the EPA sold the Superfund program to Congress with the promise that toxic sites could be cleaned up and made usable. The goal of identifying toxic sites and removing people from harm's way was never articulated. At the time the Superfund program had no experience in cleaning up a waste site, and learning how to do this was an important accomplishment of its first decade. The Superfund staff discovered that the full restoration of a toxic dumpsite was very expensive, and that often it is more cost-effective to contain the contaminants on site than to dig them up or flush them out.

The Superfund program accomplished a great deal, although advocates and the media saw it as a failure. Its only actual failure was the lack of a realistic goal. The political support needed to obtain the resources required to build a program is sometimes won, unfortunately, by exaggerating the possibility of success. When that happens it is important for the program's operating managers to redefine success and try to get buy-in from key players on the new, more realistic set of goals. Although one might argue for stretch targets and the importance of shooting high, in policy areas of great uncertainty a bit more modesty is called for when defining success. We need to give ourselves time to learn more about the problem, step back from early accomplishments, examine means and ends, and reassess the choices we have made. Politics make that task difficult, but the needs of program management make it essential.

Applying the Management Dimension of the Framework

An elegant policy design is purely symbolic without organizational capacity to perform the behaviors needed to implement the design. An environmental problem, ultimately, can only be addressed if managed organizational capacity is put into place either to control or prevent the problem. If policy makers ignore the issue of organizational capacity, then in effect they ignore the problem. In this case the analyst or practitioner knows that the policy process has not yet taken seriously the particular environmental problem.

To apply the management dimension of the framework a number of questions can be posed. Although specific questions to address vary by issue, the following questions indicate issues that need attention when applying the management dimension of the framework:

- Does the organizational capacity exist to effectively use technology that *measures* the environmental problem? If so, how much is in place and does capacity exist in the same location as the problem?
- Does the organizational capacity exist to directly utilize or encourage the use of technology or other strategic plan elements needed to *prevent* or *control* the environmental problem? If so, how much is in place and does capacity exist in the same location as the problem?
- How much experience do we have in addressing this issue or others that share its characteristics? Are standard operating procedures in place, and are they well tested and well understood? Do we know how to manage this kind of procedure, or is it something we still need to learn?
- What resources are available to develop and maintain needed organizational capacity, and are these resources adequate?
- What is the quality of the leadership in the organizations implementing this program?

The management dimension of the environmental issue is the one that tells us if the issue is considered sufficiently important to address. Although the management dimension cannot be examined without an understanding of the other dimensions of the issue, if that dimension is ignored, a symbolic response may be confused for a real one.

Next Steps

The preceding discussion illustrates the complexity of the environmental problem and, I hope, the necessity to view it from a variety of perspectives. The next several chapters seek to apply this preliminary framework to a set of environmental policy issues.

NOTES

1. In a recent revision of his book, Allison has modified his theories and added new facts for interpretation, but the basic method of analysis remains the same.

2. This is particularly important when viewing a class of issues such as environmental policy. Typically an environmental issue is seen as one of a group of connected concerns that "environmentalists" will perceive and act on as a single issue.

3. An earlier version of this section appears in a book chapter I wrote in the 1990's: "Employing Strategic Planning in Environmental Regulation," in Sheldon Kamieniecki, Richard Gonzales, and Robert Vos (eds.). *Flashpoints in Environmental Policymaking: Controversies in the 1990's*. Albany N.Y.: SUNY, 1997.

Part II

Applying the Framework

Chapter 3

Why New York City Can't Take Out the Garbage

Chapter 2 provided a framework for analyzing and understanding environmental issues. We now turn to an application of that framework for understanding the problem of disposing New York City's garbage. Examining each dimension of the city's solid waste problem will provide a comprehensive explanation of the problem and its potential solution. The city's garbage problem will be examined as an issue of values, as a political issue, and as a problem for science and technology. Finally, we will address the policy design and management dimensions of the issue.

As noted, solid waste is not only an issue for New York City but is also a national problem. In 1960 Americans generated approximately 88.1 million tons of waste per year, which is equivalent to 2.7 pounds per person every day. By 1990 that number grew to 205.2 million tons per year and 4.5 pounds per person per day. During the next decade per capita waste remained stable at 4.5 pounds per day, but total waste rose to 232 tons in 2000. From 1960 to 2000 the total amount of waste generated grew from 88 to 232 tons per year (U.S. EPA 2002a).

History of the Problem[1]

New York City's eight million residents and millions of businesses, construction projects, and nonresident employees generate as much as 36,200 tons of municipal solid waste per day. The city's

Department of Sanitation (DOS) handles nearly 13,000 tons per day of waste generated by residents, public agencies, and nonprofit corporations; private carting companies handle the remainder (DOS 2001). During the twentieth century the DOS relied on a number of landfills for garbage disposal. Then, in December 2001, the city's last garbage dump, Fresh Kills Landfill in Staten Island, closed. In response, the City Council adopted a twenty-year plan for exporting DOS-managed waste. Export became the exclusive waste disposal option for New York City (DOS 2004).

Throughout New York City's history it has had problems with solid waste management. In 1894, four years before the incorporation of the City of Greater New York, the city stopped its practice of dumping garbage into the ocean. Instead, it began a new program that included recycling and composting. Soon, however, a new administration took office, and ocean dumping resumed. A federal lawsuit filed by a group of New Jersey coastal cities forced New York City to end ocean dumping in 1935 (McCrory 1998). With plans for new incinerators slowed, first by the Great Depression and then by World War II, the city found it more and more difficult to meet its waste disposal needs. In 1947 the Fresh Kills Landfill opened. Initially the Staten Island dump was going to be a "clean fill," and the city's new mayor promised that "raw" garbage would only be landfilled at Fresh Kills for three years, the time it would take to design, site, and build a large incinerator in every borough. By the 1960s, however, one-third of the city's trash was burned in more than seventeen thousand apartment building incinerators and twenty-two municipal incinerators. The remaining residential refuse was still sent to Fresh Kills as well as to the city's other landfills (Miller 2000, xix, 233).

As we noted in an Earth Institute report: "As environmental awareness grew, public pressure began to mount against incineration and landfilling. Old landfills and incinerators were gradually shut down, with the last municipal incinerator closed in 1992." By the late 1990s only Fresh Kills remained as a waste disposal option for the residential and public waste managed by the DOS (Earth Institute and the Urban Habitat Project 2001, A-2).

In May 1996 Mayor Rudolph Giuliani and Governor George Pataki announced that Fresh Kills would receive its last ton of garbage no later than January 1, 2002. With the exception of the remains of the World Trade Center, that landfill has been closed since the last day of 2001. In an effort to determine how the city should go about disposing nearly 13,000 tons of daily waste previously sent to the site, a Fresh Kills Closure Task Force was established by the city government. The principal goal of the task force was to develop a short-term plan for diverting the waste from Fresh Kills up to its full closure in 2001. The next step was to develop a longer-term solution.

In order to divert the waste prior to closure, the city entered into a number of three-year interim contracts with private haulers. The city's annual bill for collecting and disposing residential trash jumped by nearly 50 percent, to about $658 million in 2000 and to nearly $1 billion in 2001. Whereas New York City was paying less than $50 per ton for disposal at Fresh Kills, some of the interim contracts were nearly double the price, costing more than $100 per ton when increased transportation costs are taken into account (Earth Institute and the Urban Habitat Project 2001).

In addition to the interim plan, the city developed a long-term plan to manage its waste. Under the long-term plan, approved by both the New York City Council in 2000 and the New York State Department of Environmental Conservation in February 2001, the city entered into six twenty-year contracts with private waste companies. The contracts featured fixed cost increases and, according to the Department of Sanitation, no minimum tonnage requirements. Although the plan was ostensibly for the long term, it remains vulnerable to cost escalation and increased regulation from the states that host landfills. Moreover, the plan does not include careful planning for waste transfer processes within the city. As of 2005 waste from garbage collection trucks was dumped onto the floor of waste transfer stations, where it was then loaded onto large trucks for shipment out of New York City.

In the summer of 2002 the city began to take some steps to develop elements of a true long-term plan for managing waste. While the overall waste export strategy was still being pursued, Mayor Michael Bloomberg announced a plan to develop garbage transfer stations that would compact refuse and ship it by barge for disposal. These stations would be placed in waterfront locations in each of the five New York City boroughs and would replace a system of land-based waste transfer that currently uses thousands of diesel-fueled trucks daily to haul garbage through city streets to disposal sites in other states. In late 2003 the projected expense of building these transfer stations grew, putting the plan on hold.

The current system of waste export in 2005 still leaves the city vulnerable over the long run, as both restrictions on waste disposal and its costs are likely to escalate. As landfill space continues to diminish in the eastern United States, and as political pressure from communities opposed to waste importation builds, Congress and the courts may allow states to impose restrictions on the interstate flow of municipal waste. Bills are regularly brought before Congress that would authorize local governments, state governments, and governors to restrict or prohibit the receipt of out-of-state municipal solid waste. Although the passage of such bills is far from certain, the possibility of passage over the next twenty years is substantial enough to warrant concern. Similarly stricter regulations on new landfills by federal and state Environmental Protection Agencies could increase the cost of new landfills and limit future landfill capacity. Finally, landfill operators will certainly raise prices over time, and state and municipal governments will probably enact taxes on waste disposal (Thompson 2004, 1).

NYC's Solid Waste as an Issue of Values

Our discussion of New York City's solid waste problem begins with an analysis of the values intrinsic to the issue, beginning with those that shape the consumption patterns responsible for creating 13,000 tons of residential garbage each day. The use of large amounts of packaging material, and the relatively minimal level of recycling, reflect the community's collective values. The preference for exporting waste is based on a desire to avoid the potential environmental insult of treating garbage and on the values underlying the "not in my backyard" (NIMBY) syndrome. The consumption behaviors described show little sign of fundamental change from decade to decade. Although the growth in per capita waste disposal in New York City has begun to slow, mirroring national trends, New Yorkers clearly value the benefits of a throwaway society. The value system supporting this mode of consumption dominates and has kept waste reduction off the political agenda. The issue does not pertain only to New York City; although the size and density of that city's population intensify its solid waste problem. At the root of the problem are consumption patterns that currently prevail in all modern, developed economies. New Yorkers probably place a higher value on convenience and service than may be typical, but the difference is one of degree rather than kind.

A subtle choice of values is also reflected in the way the public and the governing elite try to avoid the waste issue. Perhaps that is partly because garbage is physically unpleasant and also because it is a reminder of the great wealth some of us enjoy in the face of poverty. We discard food and clothing that could provide sustenance to the world's poor. And then, too, garbage is ugly and foul-smelling; we prefer not to think about it or where it ends up. Coupled with this attitude is the historic tendency to process garbage as far away as possible from the middle and upper classes (Bullard 1992). This helps to propagate the fantasy that all those green plastic mounds of garbage bags on the street are placed in a garbage truck and magically transported to some mythical solid waste heaven.

This pairing of convenience-driven consumption with waste avoidance is the value underpinning the city's solid waste management crisis. In contrast, throughout the twentieth century, New York invested many billions of dollars in a water supply system that is arguably the best in the world. As this demonstrates, the city does have the capacity and resources to develop long-term plans for infrastructure investment to address environmental problems—just not for garbage.

NYC's Solid Waste as a Political Issue

The value issues described in the previous section have created a climate of opinion for the politics of waste that makes it difficult for local decision makers to address the city's solid waste issues. At the core of this issue is the local politics around choosing the sites for waste disposal, transfer, and treatment facilities. Garbage is inherently undesirable, and obviously there is no positive spin that can be placed on being the host site for a community's waste. The political antipathy to waste in New York was demonstrated for more than two decades by the local politics of waste in Staten Island. The highest priority for most of Staten Island's elected officials during the 1990s was to close the Fresh Kills Landfill. In the late 1980s and early 1990s the borough engaged in an effort to secede from New York City, partly to end the use of Staten Island as the city's dump. As a sparsely populated and predominantly Republican borough in a Democratic city, Staten Island had little leverage until Republican Rudolph Giuliani was elected mayor in 1993 (Earth Institute and the Urban Habitat Project 2001, A-2).

Local politicians, with few exceptions, have caved in to the long-standing antipathy toward locating waste facilities in New York City. In the 1980s, highly conflicted but with enormous political courage, the then mayor Ed Koch obtained an agreement to site a waste incinerator in each borough. Mayor Koch's incinerator agreement collapsed during the Dinkins and Giuliani administrations, as each subsequent mayor decided that community opposition to the siting was too intense to override. Even environmentalists, while arguing against landfilling and long-distance waste transportation, are even more opposed to siting a garbage incinerator or a waste-to-energy plant within New York City's five boroughs. The politics of waste, particularly the community politics of siting, has been the principal constraint on policy options for managing the city's waste.

A number of strategies could be pursued to overcome the political opposition. A local community might accept hefty side payments in return for hosting a facility or might be receptive to hosting a waste-to-energy plant, which produces relatively clean, renewable energy through the combustion of municipal solid waste. A governor may determine that the long-term needs of the region require a permanent solution to the waste problem and develop a comprehensive region-wide program of recycling, waste reduction, waste transfer, waste-to-energy, and landfill construction. For this combination of events to occur, either the political climate would need to change dramatically or a locally elected leader would have to exert an unusual degree of political courage. This appears unlikely in the near term, and with the key political issue centered on siting, even a long-term solution to the city's waste management problem is not likely to become a serious item on the political agenda.

Exporting waste will ultimately be more expensive but, if the increased costs are gradual, they probably will not generate enough political noise to induce a sitting mayor to develop a more cost-effective, long-term plan. Similarly, although a sudden end to the supply of waste dumps poses another potential waste crisis, this, too, is unlikely to occur as long as New York City is willing to pay higher tipping fees. What is certain, however, is that any mayor who attempts to site a waste incinerator or landfill in or near the city will suffer politically.

NYC's Solid Waste as an Issue of Science and Technology

New York City's extreme population density has necessitated a number of technological innovations including an extensive mass transit network, electricity and water systems, modern sewage treatment and removal, product packaging, food refrigeration, preservatives, and, of course, solid waste removal. The technology of waste incineration has advanced dramatically since the 1960s, when much of the city's waste was burned in apartment building and municipal incinerators. Today probably the most environmentally sound methods for disposing the daily waste generated by the city's eight million residents and four million workers or visitors are regional or local waste-to-energy plants or other new and advanced waste treatment technologies. The remaining smaller volume of waste would be exported via marine waste transfer stations or rail transfer through a train tunnel from New Jersey to Brooklyn (Earth Institute and the Urban Habitat Project 2001). Marine or rail transfer of waste and modern incineration or waste treatment can significantly reduce the number of collection truck miles and also maximize the ability to collect and recycle toxic materials and heavy metals. The potential for waste leakage from landfills would also be reduced. But despite the existence of appropriate and effective waste disposal technology that is actually more affordable than the current waste system, the politics of siting continues to dominate the issue.

Experts in waste disposal are not trusted and the government lacks credibility, and so even though there is a scientific solution to the problem, politics disallows the use of the new technology. If science could reduce waste plant emissions to zero, and if experts credible to the public and interest groups could confirm the improved technology, scientific fact *might* influence the political dialogue. A non-incinerator-based technology in which garbage is treated by a closed-system chemical process and transformed into a more useful, non-waste product or lower-volume material might achieve greater public acceptance. But as demonstrated by the debate over the global climate change, the more complex the issue, the more likely that scientific uncertainty remains. Thus the NIMBY syndrome and the political self-interest of some local community-based organizations continue to take precedence over scientific progress.

Yet, at the same time that scientific technology is overshadowed in New York City's solid waste issue, technology continues to affect both the development of the problem and its potential solution. The technology of modern consumption, for example, packages food and other products in ever more environmentally damaging materials. The growth of e-commerce exacerbates this problem: compare the packaging volume to the weight of the product when Amazon.com ships a CD to your home. On the solution side, the technology of waste treatment is developing quickly and, despite my skepticism, may one day come up with a proven, clean waste treatment process that could overcome political opposition to facility siting.

NYC's Solid Waste as a Public Policy Design Issue

Until recently the use of inexpensive local landfills kept the price of waste disposal practically invisible, and so it did not pose a major fiscal dilemma. Now, however, as the expense rapidly rises, cost-benefit calculations have begun to influence the policy-making process. While the cost of disposal has risen dramatically in the past four years, alternatives to waste export have not achieved substantive status on the political agenda. If costs and regulatory obstacles continue to increase, however, the waste issue could emerge as a public policy priority, especially if the opposition of outside communities to receiving New York City waste also continues to rise. Waste disposal as a public policy issue may then be redefined, providing the legitimacy needed to find alternatives to exporting waste. At the same time, the issue will gain in significance to voters and to the politicians they elect.

The rising costs of disposal have also increased the cost-effectiveness of long-term capital investment in alternative waste disposal facilities. A result has been that waste-to-energy incineration has emerged as a possible solution to the problem, albeit remote. When New York City owned its own landfills and could dispose of waste for $20 per ton, incineration facilities were too expensive to consider. With city disposal costs already 500 percent higher than before the closing of Fresh Kills, capital investment in waste disposal technology, transfer, plants, and equipment has become a more practical solution.

Another aspect of the solid waste dilemma as a policy issue is its regulatory dimension. Local, state, and federal governments in the

United States regulate waste disposal. Individuals and apartment building staff must package and sort garbage in specified ways. If it is packaged or sorted incorrectly, fines or noncollection may result. The visibility of the issue and the immediacy of enforcement make the regulatory dimensions of this issue relatively straightforward.

One partial solution is a policy that encourages waste reduction. The tax system or command-and-control regulation can be used to reduce packaging or to encourage the development and use of biodegradable packaging. For several reasons, however, these kinds of policies would be difficult for a city government to implement. If a single city institutes stringent waste policies, large national corporations may resist compliance or simply refuse to do business in the locality, thus negatively affecting the local economy. Such rules would also require state approval, which might not be granted. These types of policies in the U.S. political system are therefore more feasible at the national level. Another option for New York City might be a fee-for-service system for waste disposal, where charges are levied for all waste pickups and rates for recycling are significantly lower than for mixed waste. However, although this type of policy design has been effective in areas dominated by single-family homes, most New York City residents live in apartment buildings, which makes it difficult to connect fees to individual behavior (Cornell Waste Management Institute 2000, 5).

Finally, the New York City Department of Sanitation itself is subject to a variety of regulations on its equipment, workforce, and waste disposal practices. Thus the DOS is both a regulator of other parties' behavior and a regulated unit itself.

NYC's Solid Waste as a Management Issue

Removing garbage from residential, institutional, and commercial locations in New York City is a major logistical and operational task. Private firms remove the waste from the city's commercial establishments, but the city's residences, governments, and nonprofit organizations produce 13,000 tons of waste each day which is removed by the DOS. To accomplish this, the department employs:

- 7,600 uniformed sanitation workers and supervisors
- 2,100 civilian and clerical workers

- 2,000 collection trucks
- 450 street sweepers
- 275 specialized collection trucks
- 280 front end loaders
- 2,360 various other support vehicles (DOS 2003)

Most of the management tasks of garbage removal do not pose a major challenge to DOS managers. Recycling, waste transfer, and final disposal are exceptions, however. Because of previous investments in specific types of collection trucks, the department must conduct separate runs for recycled paper, glass, plastic, and mixed garbage. Budget cutbacks, the high cost of separate trips, and a poor market for recycled goods forced the city to reduce its recycling program in 2002. Bowing to political pressure, the city restored the program in 2004. The social learning that had caused the gradual increase in the level of recycling was substantially disrupted by this policy change. Stopping and restarting parts of recycling confused people about what they were being asked to do, and the recycling element of DOS operations became more difficult to manage. A major operational issue in managing recycling is predicting the rate of recycled goods per household. One reason recycling costs more than traditional waste disposal in New York City is because collection trucks often return to the garage more empty than full. Because a route costs almost the same to run with full or half-full loads, the collection cost per ton of recycled waste is quite high.

Waste transfer is the process of removing garbage from the collection vehicle and transferring it to a vehicle that will bring it to final disposal. The department's long-term plan is to establish a set of marine transfer stations in each of the city's five boroughs. These facilities will allow trucks to drive in and dump their contents, which will then be compacted and loaded onto barges for shipment to a disposal site. Siting and constructing these transfer stations is a political and fiscal challenge but does not present complicated management issues. Many such facilities are now in operation worldwide. Waste transfer in New York City, even though it is well organized and successfully operated by the DOS, is currently land-based, expensive, and environmentally damaging. The key unsolved management dilemma is the price of long-term disposal and the uncertainty that funds will be available. Today the city has contracts with out-of-state landfills and incinerators to accept city waste, but the price of disposal continues to rise and the supply of disposal sites is not guaranteed.

Summary of the Multiple Dimensions of NYC's Solid Waste Issue

The politics of siting waste disposal and transfer facilities dominates all aspects of New York City's solid waste issue. This political domination limits policy design options. Some form of government-funded infrastructure is required, but the type of infrastructure that might be developed is constrained by attitudes toward waste and the politics based on those attitudes. A set of strong values and beliefs underlie the political dimension of the issue. The public's attitude toward consumption and "out-of-sight, out-of-mind" philosophy makes it difficult to combat the NIMBY mind-set. So even though science can provide data on environmental impacts that could affect policy design, the information has little chance of being incorporated into the current policy dialogue. On the positive side, New York City's government has demonstrated a high level of competence in the operational aspects of waste removal.

Studying the multiple dimensions of a policy issue clearly leads to a more complete understanding of the issue and in turn heightens insight into the processes of change. Examining the evolution of an issue provides a glimpse at levers of change that policy makers and public managers might use as they seek to shape policy. But without a strong effort at the grass-roots level, including community-based environmental education, a solution to New York City's waste disposal crisis is unlikely. If access to out-of-state waste sites becomes restricted and the price of disposal continues to grow, residents may be asked to trade off the economic costs of massive tax increases against the insult, in part symbolic, of hosting a modern waste management facility.

Conclusion

Few environmental issues are as fundamental as solid waste disposal. It is central to our health, our behavior, and how we will live in the future. Like other political issues, it has its own mythology and ideological baggage.

The waste management issue has been on and off the agenda in New York City ever since the five boroughs were consolidated as the City of New York in 1898. The current crisis of landfill capacity, on the other hand, is less than a decade old. This chapter has attempted to identify certain lessons learned from the past that may be applied in the future and in other jurisdictions. The U.S. experience provides evidence that GDP growth and increased pollution can be separated, and that a democracy can educate itself about a complex technical issue and outline workable approaches. The New York City garbage crisis illustrates the general complexity of environmental issues and affords a challenging picture of the potential for social learning.

Maintenance of the biosphere is a basic function of government within and between nations. Resolving environmental problems requires a deep understanding of specific environmental issues. Solutions must take into account the various dimensions of each issue regarding values, politics, technology, policy, and management and also consider their interactions. New York City's solid waste issue is more political than technical. Although we have the technology and management capacity to solve the problem, a cost threshold must be crossed before improved policy outweighs consumer-oriented values and the NIMBY syndrome. The need to remove waste from households is a simple matter of public health, and yet even a prosaic issue such as this is daunting to our local policy process.

The number of landfills nationwide has been reduced from more than 20,000 in the 1970s to around 2,000 today. In 2001 these landfills received approximately 55 percent of the 229.2 million tons of municipal solid waste generated (National Solid Wastes Management Association 2004).

More than five thousand communities nationwide have implemented a Pay-as-You-Throw system, making it available to 20 percent of the population. This number includes many smaller municipalities but also urban centers such as San Francisco, Seattle, Fort Worth, Austin, Buffalo, New Orleans, and the second largest U.S. city, Los Angeles, with a population of 3.8 million. As noted above, these cities have a distinct advantage in implementing this type of program as they consist largely of single-family or small-unit apartments in contrast to the high-rise residential buildings of New York City (Cornell Waste Management Institute 2000, 5).

Arguably, Chicago and San Francisco offer the greatest comparisons to New York because of their population density and prevalence of older, multi-family structures. Chicago, with a population of 2.85 million, offers residential solid waste pickup only to structures that have fewer than five housing units. Residential buildings with more than four units must contract with a private solid waste removal or recycling company. The City of Chicago, Department of Streets and Sanitation (2005) collects 1.1 million tons of residential garbage and 300,000 tons of recycling annually. All this was collected from approximately 750,000 residential structures. Thus the waste from more than 300,000 larger residential units and all commercial properties were managed by private firms (Chicago Public Library 2001).

San Francisco has a residential population of nearly 800,000. Its entire municipal waste system is contracted out to the private firm Norcal Waste Systems. Sixty percent of housing in the city is comprised of structures of five units or fewer. These smaller units participate in a Pay-As-You-Throw program through a privately contracted firm. The remaining 40 percent of the city's housing is six-units or more. The city has been in a state of flux for several years trying to implement appropriate volume/frequency rates that still provide an appropriate level of incentive for individual residents to reduce waste (DOS 2001). To its credit, San Francisco has one of the most aggressive recycling campaigns in the United States. Recycling, which is also conducted by Norcal, is available to nearly 90 percent of the city's apartment houses. The city reports a diversion rate upward of 63 percent. San Francisco has set an ambitious goal to divert 75 percent of the garbage generated by city businesses, residents, and visitors from landfills by 2010 (San Francisco Department of the Environment 2005). Clearly, other cities in the United States are more successful than New York in managing solid waste. Perhaps New York can learn from these other cities and will some day be able to take out its own garbage.

NOTE

1. This history is drawn in part from a report I co-authored entitled "Life After Fresh Kills: Moving Beyond New York City's Current Waste Management Plan" (with Gregory Frankel, Nickolas J. Themelis, and others), Columbia Earth Institute, December 2001.

Chapter 4

Why Companies Let Valuable Gasoline Leak Out of Underground Tanks

The Nature of the Problem

In Title I of the 1984 Hazardous and Solid Waste Amendments (HSWA) to the Solid Waste Disposal Act of 1976, the U.S. Congress began to regulate underground gasoline and chemical storage tanks. In 1986 the Superfund Amendments and Reauthorization Act (SARA) established a leaking underground storage tank trust fund to pay the costs of cleanups when tank owners could not be found or resisted a cleanup order from the government. Over the past century, underground storage tanks have become the preferred method of storage for the ever increasing quantities of petroleum and chemicals needed to fuel our lifestyle.

Underground tanks were used for safety and economy. They were thought to be less explosive than above-ground tanks, and land could be saved by placing them under parking lots, buildings, and other above-ground facilities. In the early 1980s, the EPA estimated that more than one-fourth of the nation's underground tanks were leaking and posed risks to the environment and public health (Feticiano 1984; U.S. EPA 1985). Around the time that tank regulation began in the 1980s groundwater contamination was becoming a political issue in parts of the United States dependent on groundwater for drinking. Releases of gasoline and its chemical additives from leaking underground storage tanks were a major source of groundwater contamination.

In 2000 groundwater consumption totaled 83.3 billion gallons per day, or about 24.1 percent of freshwater consumption (U.S. Geological Survey 2005). Although this may seem like a relatively small percentage, the EPA reported that, in 2002, 11,746 water systems serving 183.7 million people in the United States, relied on surface water, while 41,691 systems serving 84 million people relied on groundwater. Nearly all (95 percent) of our rural population depends on groundwater for daily use.

While regulation began in the mid-1980s, and much progress has been made in bringing this problem under control, the EPA reported the following figures in 2005:

- More than 670,000 underground tanks are regulated by EPA
- Since 1984, nearly 1.6 million tanks have been closed by EPA
- EPA has received reports of 447,233 tank leaks since the program began in 1984
- Cleanups have begun at 412,657 sites and have been completed at 317,405
- 129,827 cleanups have not been completed

Initially the problem of leaking tanks was puzzling, because the material in underground storage tanks was thought to be a product with economic value. In the early 1980s, it gradually became clear that toxic industrial waste creates problems when released into the environment. Toxic waste is essentially dangerous garbage and obviously of little value—the poorly understood refuse of production processes. But the material leaking out of underground tanks is the material that owners of the tanks are in the business of selling. One might assume, then, that we simply had to tell the owners that their tanks were leaking and they would want to fix them as soon as possible.

The reality of any environmental problem is complicated, and there were a variety of reasons why tanks were leaking. For one thing, underground moisture was seeping in because of leaky pipes and connections. Gasoline tanks were not protected against corrosion because no one thought they would rust. Nor did most people know that gasoline additives were harmful. Because gasoline was so inexpensive, preventing or stopping slow leaks was not cost-effective. Furthermore, many leaking tanks were the result of inattention or ignorance, such as leaks occurring at abandoned gas stations or one- or two-pump stations that were an added convenience at a rural "mom and pop" grocery store. Some leaks even occurred when gasoline was being pumped into or out of underground tanks.

As a result of tank regulation, approximately 1.5 million tanks were taken out of service in the program's first two decades (EPA 2004a). What remained, largely, were the tanks of major oil companies and their franchisees (Cohen, Kamieniecki, and Cahn 2005, 116). Consequently, there were fewer gas stations but with many more gas pumps. Self-service pumps eliminated most of the low-wage jobs associated with pumping gasoline. Moreover, pay-at-the-pump technology a few years later eliminated many of the cashier jobs that had survived the advent of the self-service pump. These changes resulted in an ever increasing number of underground tanks and a shrinking number of people to keep track of their contents.

Environmental damage to the property and health of people living near leaking tanks is an important dimension of the problem. It became clearer in the 1980s and 1990s that the owners and operators of leaking tanks were liable for the costs of these damages. The potential and actual liability of leaking tanks, along with increased government regulation of tank installations, resulted in tank owners and operators changing their behavior. There were also many closures and sometimes abandonment of existing tanks. Given all the existing and potential costs of tank leaks, it is difficult to understand why more than four hundred thousand tank leaks have been reported in the past two decades (U.S. EPA 2002c).

Leaking Tanks as an Issue of Values

The development of the problem of leaking underground storage tanks has several value dimensions. The value of human safety and protection of property was what stimulated the burial of tanks in the first place. Many tanks were installed during the 1950s and 1960s, a time when our knowledge of ecology was limited, and placing tanks underground should be seen largely as an expression of ignorance and not a lack of concern about protecting groundwater. The second relevant values are mobility and economic consumption. We built gas stations everywhere to give us maximum freedom to express our economic preferences. These values also contributed to the expansion of suburban development and sprawl, which, in turn, created demand for the placement of underground storage tanks throughout the countryside (Cohen, Kamieniecki, and Cahn 2005, 116).

What do I mean by the values of consumption and mobility? The huge consumption of material goods and services by American society is well documented and need not be reiterated here. With regard to the latter value, it is undeniable that America is a mobile society. Approximately 43 million people, or 16 percent of all Americans, relocate each year (U.S. Census Bureau 2002, section 3–1). When we relocate we expect certain distinct features in our new location as well as a familiar set of goods and services. Our desire for mobility is also reflected in travel data. According to the American Automobile Association (2004), more than 38 million Americans traveled fifty miles or more during the July 4 weekend in 2003. Mobility in the United States is almost synonymous with freedom. The extreme value placed on mobility in American culture requires expending a significant amount of energy. Coupled with the value placed on owning one's own home in a rural-like setting, the resulting suburban sprawl has placed gas stations in remote and ecologically fragile places. The problem of leaking tanks, therefore, is a result of consumption patterns resulting from these value choices.

In a free market economy, where production and consumption are highly valued, both are seen as needing no justification. Even environmental protection is justified as a form of consumption. We need to protect the environment so we can enjoy it. Gasoline stations are placed wherever there is sufficient demand for gasoline. The impact of the development and travel that gasoline enables is an effect to be mitigated if necessary. So great is the priority placed on the value of consumption that it is rarely questioned.

As a willing participant in the American system of freedom, mobility, and high energy consumption, I am not arguing for its end but simply pointing out its unquestioned acceptance. We could live according to a set of values that placed a higher priority on preserving ecology or that promoted a spiritual rather than material lifestyle. An expression of the value of ecological preservation would be a policy that required greater development planning and a more meaningful assessment of environmental impacts than currently exists. Although the policy we now have does require that we consider environmental impacts, it also allows us to ignore those effects.

Appreciating the values involved in the problem helps us to understand the presence of underground tanks throughout the United States, including in environmentally sensitive areas. They do not explain, however, why the tanks leak. While a throw-away society clearly discards certain valuable resources such as gasoline, it does not advocate the disposal of unused products of value such as gasoline. To understand that aspect of the problem we need to look at the issue through other lenses.

Leaking Tanks as a Political Issue

In the early 1980s we discovered that a substantial proportion of the nation's underground tanks were leaking. Politics did not cause the tanks to leak, of course, but preventing future leaks and cleaning up the damage of past and existing leaks rapidly became a political issue. The market value of gasoline was not high enough to ensure self-regulation. Oil companies and tank owners allowed tanks to leak and did little to prevent or clean up tank seepage. If tank leaks had no impact outside the tank owner's property, the issue might have stayed off the political agenda. But because petroleum is easily transported through the ground, it can readily move off the tank owner's property to contaminate other peoples' property and groundwater.

With the contamination of individual property and collective resources, and the failure to develop a private response to those problems, leaking underground tanks became a political issue. The problem had three dimensions that reached the political agenda: the development of standards for new and existing tanks to prevent future leaks; clarification of the specific liability incurred by tank owners for leaks; and methods of assigning responsibility for the payment and execution of cleaning up the damage.

The issue of underground tanks arose amidst the consolidation of the U.S. oil industry and a change in the distribution system for gasoline. Small mom-and-pop grocery stores and auto service stations that also had gas pumps were being replaced by larger gas stations with more pumps and tanks and no auto repair shops. In the mid-1990s these gas stations began to include twenty-four-hour convenience stores. Tank regulation contributed to these trends by increasing the regulatory requirements for underground tanks and thus operational costs as well. As a result, increased costs and complexity drove small operators out of the retail gasoline business.

The politics of tank regulation was relatively low-key by environmental standards. Interest groups representing the oil industry initially and reflexively opposed government regulation in this area when it was first proposed in the early 1980s. Petroleum was excluded from the Superfund's stringent joint, strict, and several liability provisions, and the industry wished to maintain that exclusion (Cohen, Kamieniecki, and Cahn 2005, 164). Eventually the industry came to support some form of tank regulation. Because gasoline contains substances besides petroleum, the Superfund liability provisions could be applied to impacts from other toxic additives, particularly benzene, lead in older tanks, and, in the newer tanks, methyl tertiary butyl ether (MTBE), the antiknock replacement for lead in gasoline. The larger oil companies also saw a benefit in improving the management of their product. By raising the costs of staying in the business, they were able to consolidate the distribution system under their control. When corporate leaders became aware of the size of the leaking tank problem, their posture switched from opposition to a desire to influence and moderate tank regulation. It became clear that some form of tank regulation would simply become part of the cost of doing business in the United States.

While large oil companies saw tank regulation as a necessary evil, the EPA had to be sensitive to the impacts on smaller businesses of excessively stringent or costly regulations. The final regulation governing tank management took fourteen years to finalize and was not issued until 1998. The sheer number of regulated parties created a complex regulatory environment.[1] Unlike the oil industry, which is dominated by a small number of large companies, gasoline distribution includes a large variety of smaller players and franchisees. In this respect, tank regulation could be seen as a subset of the general field of government relations with small businesses. In American political mythology, small business is considered a critical point in the progression from rags to riches and a key part of the American dream. In this scheme, the small business owner is a former worker who has managed, through hard work and good fortune, to become upwardly mobile. With this image in mind, it is politically dangerous for government to treat the small grocery store owner who has a gas pump out front the same way it treats a major oil company. Nonetheless, in the 1980s and 1990s, the gasoline distribution industry was consolidated and tank regulation was combined with a range of other factors that contributed to the demise of many small operators. Tank regulation probably escaped blame for this trend.

The tank issue, of course, is not simply about regulating tanks to prevent leaks; it also involves the cleanup of leaks. Gasoline from a leaking tank is no longer a valuable product but instead is a potential toxic waste. The politics of underground tanks therefore becomes the politics of hazardous waste cleanup. The effect of hazardous waste tends to be severe and extremely localized, making it one of the most volatile and unpredictable issues in environmental politics. The American political system is custom-made to address such issues. With its single-member, winner-take-all districts and its federal and highly decentralized structure, the system responds best to issues with clear, geographic origins and impacts. Representatives to Congress, to state and county legislatures, and to town or city councils are expected to respond to demands for action on issues such as these. The geographic orientation of American politics is well known and accepted, and so a range of rules and customs have been instituted to help politicians respond to the needs of their constituency base. The American tendency toward ad-hoc, nonpartisan, issue-based politics has also led to the creation of hundreds of local organizations lobbying for action on toxic cleanup.

While tank regulation is a low-intensity and relatively noncontentious political issue, the cleanup of tank leaks is a different matter. The politics of hazardous waste cleanup is examined in greater detail in chapter 5, but one point is important to raise here, and that is the relationship between the politics of toxic waste and the sheer complexity and unpredictability of the threat it poses. The chemicals in toxic waste vary widely, as do the pathways leading to human exposure. A leaking tank in one location can quickly contaminate the well that supplies drinking water to a nearby home or a neighbor's basement. In a different location, on the other hand, the leak can go undetected and be contained for decades. One reaction to this complexity is fear and an understandable refusal to gauge the actual risks that toxic releases pose. This leads to a highly intense, often symbolic and contentious political dialogue. One reason is that many of the health effects of toxic substances have a long latency period, and so the absence of an immediate health effect does not guarantee that the waste is not potentially hazardous to one's health in the future.

The degree to which the politics of tank regulation evolves into the politics of hazardous waste is unpredictable, but it is part of the political side of the tank issue. Although tank policy and management is not typically high on the political agenda, the hazardous waste element involved reminds us that sometimes such an issue can develop a highly volatile political dimension.

Leaking Tanks as an Issue of Science and Technology

Do we have the technology to prevent leaking tanks? We most certainly do. Do we have the technology to clean up leaks from tanks? Not entirely. But we have learned a great deal in the past quarter-century about how to engineer a cleanup. And while the technology to prevent leaks is more readily available than the technology to clean them up, overall I would not characterize leaking tanks as a technical challenge. One might conclude that the problem of leaking tanks was largely caused by the development of the automobile and its effect on our patterns of land use. Still, the technical know-how exists to transport, store, and pump gasoline while minimizing hazardous releases into the environment. Tanks can be built with double walls to minimize the probability of leaks, and leak detection devices can be installed on tanks to quickly determine if the tank's integrity has been breached.

Once a tank has leaked, however, we have the technology to remove the toxins only from certain environments. Some environments make it difficult to detect the presence of gasoline and its additives in the ground. In certain cases it is difficult to actually remove the pollutants from the environment. The common techniques used for toxic waste cleanups here are containing the waste in one area, pumping water through the site while filtering the toxics, and, where necessary, removing the soil and placing it in a designated toxic waste dump. These tasks have been performed before and have varied levels of effectiveness depending on the waste, the local ecology, and the quality of the equipment used to treat the waste (Blackman 2001, 142–189).

Leaking Tanks as a Public Policy Design Issue

The tank problem, in some respects, is a result of the relatively low price of gasoline in the United States. It is sometimes more cost-effective in the short run to let a tank leak than to stop the leak. Effective environmental policy provides incentives for institutions and individuals to alter the behaviors that are contributing to environmental damage. For example, tradable air emission allowances encourage businesses to reduce emissions in order to sell allowances to other firms for cash. What behaviors need to change to resolve the problem of leaking tanks? The efforts early on to create a pattern of decentralized development, which resulted in locating gasoline stations near underground water supplies, led directly to the problem, and it is far too late to influence those behaviors. The issue facing policy designers today is to ensure that the gasoline in underground tanks, pumps, pipes, hoses, automobiles, and gas tanks of trucks stays where it belongs. The desired behavior, when tanks leak, is for the responsible parties to clean up the leak and pay their neighbors for damages.

The EPA approached the problem by working closely with industry and with state and local governments on several initiatives. First, tank standards were set that reduced the probability of leaks in new tanks. Second, tank owners were required to implement a method to detect leaks. And, third, tank owners had to demonstrate that they had the financial capacity to clean up leaking tanks. In the twenty years since tank regulation began, the United States has made significant progress in reducing environmental threats from leaking tanks, as will be demonstrated below.

The Underground Storage Tank (UST) Program has two key indicators of success: the number of tanks leaking and the number of leaks cleaned up. According to EPA estimates, approximately forty-six thousand tanks leaked each year prior to regulation in the early 1980s. Since the inception of the EPA Office of Underground Storage Tanks (OUST), the average annual number of confirmed leaks has been thirty thousand, with fewer over the last few years (U.S. EPA 2002b).

Since 1988, state agencies reported significant results in tank closures and cleanups, with approximately 10,000 to 40,000 cleanups completed each year. As of March 2004, OUST reported that 1.6 million unsafe tanks had been closed since its inception. More than 440,000

UST leaks have also been confirmed. Of this number, more than 317,000 cleanups have taken place and 129,000 cleanups remain (U.S. EPA 2005a). Despite this progress, the problem of leaking tanks persists. The rate of tank leaks has declined, but the problem remains unsolved.

Here we have private parties working together with all levels of government, a policy design that appears to be successful, and apparent progress in solving a societal problem. In short, economics, politics, and environmentalism are in reasonably good alignment, and yet the problem persists. We can solve this puzzle, however, by looking at the problem of leaking tanks as one of organizational management.

Leaking Tanks as a Management Issue

In the late 1980s I attended a meeting chaired by Ron Brand, the first director of OUST, on approaches to leak detection at gasoline stations. One staff person spoke about inventory control as a possible method. The idea was simple: a gas station attendant simply had to read the dial on the pump each day to see how much gasoline had been pumped from the tank and then compare it to the amount of gasoline that had been delivered. To verify the result, a dipstick could be used to measure how much gasoline remained in the tank. In response to this suggestion, Ron described the following scenario: "It's 6:00 in the morning in February, and you work in a gas station in Montana where the wind chill this morning is –34 degrees. What are the chances you'll actually go outside to do anything other than pump gas? In all likelihood, your idea of inventory control is a work of fiction." Demonstrating his keen insight into how organizations function in the real world, Ron pointed out that what looks simple to a policy designer may be entirely unrealistic.

A major problem in regulating underground tanks is the sheer number of them and the multiple actions required to ensure compliance. Small leaks that result from overfilling tanks are one example. If you pump your own gas at a filling station—a behavior encouraged by self-service pumps with a credit card option—you'll notice a sign with this direction: "Do not top off your tank. Please stop pumping when the automatic cut-off engages." A former practice at gas stations was to fill the tank with as much as it could hold, which often resulted in gasoline ending up on the ground. Probably millions of people disregard those instructions every day and overfill their tanks. Although these are obviously very minor leaks, they add up.

The behavior of millions of people needs to change in order to prevent leaks. Among them are people who manufacture, deliver, and pump gasoline; build, install, and repair tanks; drive vehicles and pump their own gas; and insure, inspect, and regulate tanks. The organizational capacity to do this work has been evolving for more than seventy-five years, with many additional new tasks and routines over the past decade. The amount of work involved in these tasks continues to grow, along with the number of miles traveled by vehicles in the United States. The complexity of this work has also increased with the advance in technology and the implementation of many more regulations. The reason tanks continue to leak, in my view, may be found within the vast scope of the effort to distribute and use gasoline for transportation. First, the sheer number of transactions means that there are numerous places within this chain of production for mistakes to occur. Second, some tasks such as leak detection are new, and the organizational capacity to perform them is still in development.

It may seem that the management of these prosaic activities ought to be simple enough to make leaks less common. Case studies of new tank leaks reveal that most leaks are apparently caused by human error that could have been prevented, such as damages that occur during installation or connections between pipes and tanks coming loose (Blackman 2001, 378). Leaks are also sometimes caused by extreme weather conditions. In all cases, the damage can be minimized by detecting leaks early, which requires tank owners to develop the capacity to find and stop leaks quickly. This involves hiring an adequate staff, purchasing needed equipment, developing standard operating procedures, training staff in those procedures and in the use of equipment, and, finally, implementing the procedures. In other words, it is a matter of fundamental management.

A Summary of the Many Dimensions of Leaking Tanks

The relatively high number of leaks must be seen, in my view, as generally a management issue. The policy now in place has resulted in a reduction in the number of tank leaks. Leaking the product into the ground instead of selling it obviously has no economic benefit, and there is little private-sector lobbying to ease the rules on tanks. Arguably, then, if gasoline were more expensive, people would be more motivated to prevent leaks. This may be true, but when the potential liability tank owners face from third-party damages resulting from leaks is added to the value of gasoline, it is hard to see the tank problem as one of poor regulatory design. Tank owners have plenty of motivation to behave "correctly." Perhaps they simply lack the capacity to do what is in their obvious self-interest.

The number of transactions and behaviors required to protect the environment exacerbates the management issue at the heart of the underground tank dilemma. The fact is that we have chosen to remain a highly mobile society dependent on a specific technology, namely, the internal combustion engine, which requires gasoline, a toxic substance, to run. The management question is this: can we make every driver, every person delivering gas, every gas station attendant aware of the urgency to handle gasoline very carefully? Can we perfect the process of installing tanks and pumps to reduce the probability of leakage? Is it possible to develop a level of skill and technology that prevents most leaks? Can our organizations that extract, refine, deliver, and distribute oil and gasoline develop the capacity to reduce leaks? The answer to most of these questions is yes, as evidenced by the significant reduction in the number of leaks in the past two decades. But a question remains: How much more reduction is necessary and feasible?

In terms of the science and technology dimension of the problem as defined above, do we know how to prevent leaks? The answer is clearly yes. Our inability to develop the organizational capacity to deploy this technology, however, requires that we redefine the technological issue. The question then becomes, do we know how to run vehicles without using toxic substances as fuel? Here the answer is less clear. While electric and hybrid cars do exist and many experiments have been performed with vehicles using alternate fuels, as of 2005 this technology is hardly proven and certainly not yet widely accepted in the marketplace.

Therefore, although the problem of leaking tanks is clearly an issue of organizational capacity—that is, the ability to control individual and group behaviors to reduce the frequency of oil leaks—the solution may lie in the development of a technology that can be used safely even without a high level of organizational capacity. One problem with issues of organizational capacity is that people who know nothing about the problem often assume that a solution can be easily or automatically developed. If the policy design makes sense, they reason, and the politics looks favorable, then the policy can virtually implement itself. Frequently management complexity is underestimated and organizational capacity assumed. The fact is that nothing is self-implementing. People must perform certain tasks, working together in an organizational setting. And if they are to avoid making mistakes, the system must include a high level of expertise and training.

Consider the example of nuclear fuel and waste. The risks of contamination and of illegal sale of fuel to terrorists require that those who manage these materials are extraordinarily skilled and committed to having the resources needed to avoid mistakes. The risk of failure is potentially catastrophic, and its probability at some point is quite high. As we have seen, the safe handling of these materials is most often a management problem, and yet the level of proficiency needed to manage the problem may be beyond human capacity. The issue of nuclear materials is so profound that their very use became a political matter. Ultimately both political and economic factors brought an end to the construction of nuclear power plants in the United States. In effect, the managerial dimension in this case became political. Because it seemed that we were unable to manage nuclear waste, political opponents to nuclear power used the management issue to defeat the use of that power.

Gasoline is obviously not as dangerous as nuclear material, but its use is so widespread that the potential danger it poses, while not as intense in a single location, has a far greater scope. In sum, the issue of leaking tanks can be reduced to three questions: Can we reduce the number of leaking tanks? Yes, we already have. Can we develop a management system that can prevent a leak from ever occurring? Probably not. And what is the minimal risk level we can endure and the pace of cleanup we need to maintain the quality of our groundwater and environment? We do not know.

Conclusions

The problem of leaking underground tanks, as defined by the political and policy process in the United States, is principally a management problem. Tank legislation and rules have broad political support, and the EPA's approach to implementation has been strategic and effective in reducing the numbers of tank leaks in the past. This reduction of leaks has not resulted in an end to the problem, however, and its persistence should be a cause of real concern. When toxic substances damage aquifers, the damage can be irreversible. Either the technology to clean up the aquifer does not exist, or its use would be so expensive that, for all practical purposes, the technology is unavailable. Underground tank leaks aside, however, we certainly have the technology to transport and pump gasoline without spilling it.

But can we operate and maintain that technology without releasing gasoline into the environment? While we may have problems getting organized for those tasks, at a deeper level the tank problem can be seen as having been created by the technology of the motor vehicle. It would be easy to argue that we are stuck with the automobile culture and its resultant patterns of land use development in the United States and, to a lesser degree, in other developed countries, and, of course, we are. However, we can keep our automobiles and maintain our current lifestyle while changing the fuel that automobiles use. The problem of underground tanks can be truly resolved when we no longer need as many of them.

Leaking underground tanks are a good example of an environmental issue that is easier to understand through the multidimensional framework presented in this book. The value of mobility led to the problem of leaking tanks. It is clearly in the economic self-interest of the regulated community to keep this product from leaking. The failure to prevent leaks is mainly a management issue, but the issue is whether we can simplify the mass behaviors required to manage this problem and reduce the number of errors common to the current system of gasoline distribution. While it may be premature to propose a management solution to this problem, the management problem could be unsolvable. If it is, then the technology that fuels our mobility (excuse the pun) needs to change if we are indeed to solve this problem.

NOTE

1. As of March 2004, approximately 680,000 active underground storage tanks (USTs) were subject to UST technical regulations (U.S. EPA 2004b).

Chapter 5

Have We Learned How to Clean Up Toxic Waste Sites, and Can We Afford It?

The Nature of the Problem

The reader may notice that the chapters in this section are getting increasingly longer, and it is not an accident. The environmental problems we are seeking to understand grow in complexity as this volume progresses. The issue of toxic waste creation and cleanup is multidimensional and quite intricate. As I noted in a book chapter I wrote in the early 1980s:

> Every time we wrap a slice of cheese in plastic wrap, purchase a nylon backpack, or drink coffee from a Styrofoam cup, we are contributing to this nation's hazardous waste problem. We all use and benefit from goods that, when manufactured, create toxic wastes as unused by-products. The production of plastics and synthetic chemicals increased dramatically after World War II. Goods previously made of wood, wool, and cotton were replaced by those made of plastic, polyester, and similar substances. These synthetic materials lasted longer than "natural" materials, and for a time were less expensive. (Cohen 1984, 275)

As a result of these consumption patterns and our methods of disposing the by-products of our production processes at that time, I also made the following observation:

> A huge amount of hazardous waste already permeates our ground and water, and the amount being generated now is staggering. In

> 1980 EPA estimated that Americans produced more than 57 million metric tons of hazardous waste per year.... In 1980 EPA estimated that approximately 90 percent of the nation's hazardous waste was disposed of by environmentally unsound methods. (Ibid., 275)

EPA's early data indicated that there were between thirty thousand and fifty thousand hazardous waste sites in the United States. We knew neither their locations nor the risks they posed. Toxic waste did not become a critical issue until the late 1970s, when a series of incidents revealed that communities were at risk from wastes they had either been unaware of or had assumed were safely disposed of through burial or other methods. The most dramatic example was the toxic waste site at Love Canal, near Niagara Falls, New York. This case is worth recalling, because in many respects it illustrated all the dimensions of the toxic waste problem. Love Canal was an abandoned canal that from 1942 to 1953 was used as a municipal and chemical dumpsite. According to the on-line archives at the State University of New York (SUNY) at Buffalo's Love Canal Collection:

> In 1953, with the landfill at maximum capacity, Hooker [Chemical Corporation] filled the site with layers of dirt. As the post-war housing and baby boom spread to the southeast section of the city; the Niagara Falls Board of Education purchased the Love Canal land from Hooker Chemical for one dollar. Included in the deed transfer was a "warning" of the chemical wastes buried on the property and a disclaimer absolving Hooker of any further liability.
>
> Single-family housing surrounded the Love Canal site. As the population grew, the 99th Street School was built directly on the former landfill. At the time, homeowners were not warned or provided information of potential hazards associated with locating close to the former landfill site.
>
> According to residents who lived in the area, from the late 1950s through the early 1970s repeated complaints of odors and "substances" surfacing in their yards brought City officials to visit the neighborhood. (ETFNF 1998)

In the spring of 1978 a thaw after an unusually snowy winter (even for western New York) resulted in the partial collapse of the schoolyard at the 99th Street School and the unearthing of a number of barrels containing toxic waste. One mother living in this neighborhood, Lois Gibbs, was trying to raise her family and found that her son was ill. Ms. Gibbs claimed that her son's illness resulted from the effects of the dumpsite. According to Gibbs:

> After reading the newspaper article about Love Canal in the spring of 1978, I became concerned about the health of my son, who was in kindergarten at the 99th Street School. Since moving into our house on 101st Street, my son, Michael, had been constantly ill. I came to believe that the school and playground were making him sick. Consequently, I asked the school board to transfer Michael to another public school, and they refused, stating that "such a transfer would set a bad precedent."
>
> Receiving no help from the school board, city, or state representatives, I began going door to door with a petition to shut down the 99th Street School. The petition, I believed, would pressure the school board into investigating the chemical exposure risks to children and possibly even into closing the school. It became apparent, after only a few blocks of door knocking, that the entire neighborhood was sick. Men, women, and children suffered from many conditions—cancer, miscarriages, stillbirths, birth defects, and urinary tract diseases. The petition drive generated news coverage and helped residents come to the realization that a serious problem existed. (Gibbs 1998)

In that same year the New York State Department of Health collected and analyzed air and soil samples in more than two hundred homes near the abandoned canal and conducted health tests on the families living in those homes. According to the SUNY archives:

> On August 2, 1978, the New York State Commissioner of Health, Robert M. Whalen, M.D. declared a medical State of Emergency at Love Canal and ordered the immediate closure of the 99th Street School. Immediate cleanup plans were initiated and recommendations to move were made for pregnant women and children under two who lived in the immediate surrounding area of the Love Canal.
>
> The president of the United States, Jimmy Carter, declared the Love Canal area a federal emergency on August 7, 1978. This declaration

would provide funds to permanently relocate the 239 families living in the first two rows of homes encircling the landfill. (Ibid.)

The problem with this first response was that while the two blocks surrounding the canal were provided help, the next ten blocks of the same neighborhood adjacent to the dump were not included in the response. Lois Gibbs and her family lived three blocks from the canal. Gibbs organized the Love Canal Homeowners Association and led its efforts to demand a more comprehensive response to the contamination of the neighborhood (Levine 1982, 44–45). Her skill as an advocate eventually took her to Washington D.C., where she became a national leader for the cause of cleaning up toxic waste sites. In the decades following Love Canal's "discovery," this talented and motivated "average citizen" successfully organized hundreds of communities to protest and demand the cleanup of toxic waste sites. It turned out that Love Canal was not unique. Thousands of areas throughout the United States have had toxic waste either stored or dumped there, and residents and workers in the near vicinity were not aware of it. Much of the waste was buried and, after years of neglect, was seeping underground into aquifers, basements, and any other location where water and gravity could carry it. When Superfund, the first federal program for cleaning up toxic waste, was enacted in December 1980, we were only beginning to understand the problem of toxic waste sites and spills (Cohen 1984, 275). A quarter-century later we have a greater understanding of the dimensions of the toxic waste problem and have slowly begun to make progress in cleaning it up.

The Early Response: The Superfund Program

By the end of 2002 the EPA had assessed the contamination levels of 44,418 sites. Of these sites, 33,106 had been found to present levels of risk low enough that redevelopment was permitted. The remaining 11,312 were either in the assessment stage or on EPA's list of priority sites otherwise known as the National Priorities List (NPL). In addition to the long-term cleanup of toxic waste sites, Superfund also includes a program of emergency response cleanups, called "threat removal actions." Since 1980 Superfund's threat removal program has conducted more than 7,400 removal actions (U.S. EPA 2003c).

Waste sites that pose the greatest threat to human health and environmental quality are placed on the NPL. As of late 2002, 1,560 sites had been proposed, officially added to the list, or deleted from the list. Of those sites:

- 61 were proposed and being reviewed for inclusion on the NPL.
- 1,499 sites were either final or deleted from list.
- Of the 1,499 final and deleted NPL sites, 850 were considered under control, that is, they had either been deleted or the construction of the cleanup solution had been completed.
- 382 sites had remedial construction under way.
- At 248 sites, the nature of the contamination issue was being analyzed or cleanup strategies were being designed.
- 19 sites had not begun either study or design work.

Of the sites where construction was "completed," 589 had what EPA calls "post-construction activity under way" (U.S. EPA 2003c), which includes applying a variety of techniques to determine whether contamination has escaped the site. To oversimplify, the three basic ways to clean up a site, briefly outlined in chapter 3, include (1) removing the contaminated soil and bringing it to another location either to store or transform it to a nonhazardous material; (2) flushing water through the site and filtering the toxics out of the site over a long period; and (3) building a structure around the site and a cap on its top to contain the toxics (U.S. EPA 2003c).

The term "Superfund" was derived from the method of financing the cleanups. The amount of funding on toxic waste cleanup has varied but in the last decade has been institutionalized at a little more than $1 billion per year. Originally, in 1980, the Superfund included funds from both general revenues and a dedicated tax on oil and chemical feedstock. Funding authorization in the Superfund grew from not more than $210 million per year in the early 1980s to about $1.4 billion per year between fiscal year (FY) 1987 and 1990. In FY 1991–93, funding authorization peaked at about $1.6 billion (U.S. EPA 2005b). According to the Congressional Research Service, appropriations—the money actually provided to EPA for spending—grew from $40.3 million in FY 1981 to $1.5 billion in FY1997. From 1981 to 1997 Congress appropriated a total of $17.9 billion toward the

Superfund (Reisch and Bearden 1997). By the mid-1990s Superfund spending ranged from $1.3 billion to $1.7 billion per year. For FY2004, the Bush administration asked for $1.39 billion for the program, an increase over the $1.27 billion appropriated for FY2003 (U.S. EPA 2005b). Bearing in mind that these are actual dollars and do not reflect the lost value owing to inflation, it is important to note that actual spending peaked in the mid-1990s. In 1995 the Superfund's dedicated taxes expired and neither Presidents Clinton or Bush requested reauthorization of the tax. Instead, funds were allocated from general tax revenues (Pianin 2003). Although substantial funding is allocated to waste cleanup, many environmental advocate groups believe that the funds allocated are insufficient and the pace of cleanup too slow (Switzer and Bryner 1998, 102; Hird 1994, 204).

EPA's enforcement program, through actual and threatened lawsuits, has compelled private firms to spend an estimated $20 billion on cleanup activities (U.S. Government Accounting Office 2003). It has been difficult to estimate total government spending on cleanup as opposed to other costs of the Superfund program, which is especially true because of the controversy stemming from the high costs of overhead in the program. Superfund spending dwarfed all other EPA programs during the 1990s. Many in the private sector have complained that more Superfund dollars were spent on planning, study, and lawyers than on actual cleanup operations. Regardless of how much money has been wasted, at least $100 billion in public and private resources has been spent on toxic waste cleanup in the United States over the past quarter-century (Nakamura and Church 2003, 9).

A great deal of time, resources, and effort have clearly been expended cleaning up the mistakes of the past. How did we get ourselves into this expensive and dangerous mess? Part of the answer is ignorance, which is discussed further in the section on values that follows. But for now it is sufficient to say that decision makers believed that burying waste underground would reduce the danger caused by that waste. Regulation of hazardous waste did not begin until the Resource Conservation and Recovery Act (RCRA) of 1976 created what came to be called "cradle to grave" regulation of hazardous waste. Although enacted in the mid-1970s, this regulation was not operational until the late 1980s. RCRA set standards for hazardous and nonhazardous landfills and for the transport, storage, and disposal of all waste. A manifest system to track all hazardous waste was a critical part of this traditional command-and-control regulatory system. The act required that every shipment of waste be accompanied by a detailed inventory report (O'Leary et al. 1999, 33).

While the goal of RCRA was to prevent the creation of new Love Canals, the United States still had to deal with the legacy of sloppy hazardous waste management from the dawn of the industrial age. Even with a new regulatory system in place by the mid 1980s, the United States continues to face the problem of midnight or illegal dumping of hazardous waste. Because the management of hazardous materials is an expensive matter, unscrupulous manufacturers can cut their costs by handing their waste to illegal operators (Edwards, Edwards, and Fields 1996, 39).

RCRA has been successful in reducing the amount of hazardous waste produced and discarded in the United States. As the EPA noted in its report, *25 Years of RCRA: Building on Our Past to Protect Our Future* (2002d, 6):

> In 1980, nearly 50,000 businesses generated hazardous waste, and about 30,000 businesses ran waste treatment, storage, or disposal facilities (TSDFs). In 1999, only 20,000 businesses produced hazardous waste, with about 2,000 TSDFs managing that waste. What's more, the amount of hazardous waste disposed of in landfills has gone from 3 million tons to less than half that amount.

Of course, as we saw in chapter 3, despite a similar reduction in the number of USTs and even a substantial reduction in tank leaks, leaks persisted and continued to damage the environment. Of course, gasoline leaks from USTs appear almost benign when compared to the toxic chemical cocktail present in most toxic waste dumpsites. Although the *pace* of creating new Love Canals has probably been reduced, new hazardous sites have continued to emerge over the quarter-century since the enactment of RCRA. While clearly RCRA has contributed to the slower pace at which toxic waste sites are created, de-industrialization in the United States has also been a factor. The toxic waste we once created when making plastics has very likely gone to the developing world along with the factories that produce the plastic (Clapp 2001, 104).

Toxic Waste as an Issue of Values

Why did we create a toxic waste problem, and in what respects does this problem reflect our individual and collective values? Why did we allow businesses to mishandle this extremely hazardous waste, and why were we not aware that the waste was dangerous? We can begin to answer some of these questions by examining the value system underlying our response to the wastes produced by our system of production and consumption, we can begin to answer these questions.

Deeply embedded in American culture and myth is the importance of private property, privacy, and individual freedom. Americans tend to regard any interference in their use of their own property as infringing on their rights. We speak of the sanctity of private property and the need to limit the government's role in telling us how to use our property (Echeverria and Eby 1995, 2). Our view of land as private property is part of this mind-set. We are told that Native Americans, having a different set of values than ours, did not originally have a concept of land as property, and only learned gradually and painfully of the impact of this notion. The early proponents of the environmental movement, Rachel Carson and Barry Commoner, focused much attention on the need to demonstrate that what we consider discrete "pieces" of property are, in fact, interdependent components of ecological systems. Carson demonstrated the movement of toxins through the biosphere, and Commoner succinctly noted that "everything is connected to everything" (1996 [1971], 163).

Our property laws recognize that if a fire in a neighbor's home spreads to your house, the neighbor is responsible to reimburse you for damages. But what if the damage is caused by an air pollutant created five hundred miles away? What if the culprit was a chemical buried a mile away and was carried by an aquifer or underground stream into your backyard? The freedom to act as you wish on your own property assumes that the impact of your behavior will be limited to that property. The problem of toxic waste is not restricted to the damage caused on the land under the dumpsite; it is also the harm that is caused when toxics migrate off-site. The concept that "private property" is truly private, never an absolute in Western law, is significantly compromised by our dependence on and awareness of ecology.

We came to understand the importance of ecology just recently and began to integrate it into our value system only in the late twentieth and early twenty-first century. Before then, however, the values of ownership and private property were strongly held, and it was not generally understood that toxic chemicals released on one's property could have dangerous impacts outside that property. In other words, it was not regarded as a problem if private activities included depositing toxic waste on the ground or in the water, as long as the dumping ground legally belonged to the dumper. Scientists working for companies dumping these materials undoubtedly knew, or suspected, that such materials could be hazardous (Levine 1982, 11–12). But most management personnel in the early and mid-twentieth century were as ignorant of these issues as the political leaders of the time.

Belief in the principles of free enterprise and capitalism represent a second and related set of values. According to this ideology, free enterprise creates wealth, which in turn creates a higher quality of life. Private enterprise is built on individual initiative, and government intervention only slows individual and collective progress toward greater wealth. According to this view, the government that governs least governs best. Collective community action is reserved for soccer leagues and bake sales to buy new computers for a local school. The police function is reserved for individual misbehavior. Under this value system we assume that corporations are motivated by the wholesome and creative desire to make a profit.

Here I digress for a moment to note that while I favor free enterprise, I also believe that, over the long term, it is compatible with environmental protection. The capitalist system has generated a material way of life in developed nations that has facilitated human comfort and intellectual and social progress. Home ownership is one example of the power of free enterprise in the United States. According to the U.S. Census Bureau (2003), nearly 68 percent of Americans live in private homes. The pride of ownership inspires people to make improvements in their homes, and, in doing so, they build a stake in the society and its political and social stability. This great unleashing of private ownership would never have happened, however, without the innovation of the guaranteed or government-insured mortgage. Before this type of government intervention, the average person could not afford to make a down payment on a home. Without a strong government role, a private firm would probably not have taken the risk to loan 90 percent of the value of a house with the house itself as collateral. It is not the job of private firms to risk their capital in that way.

But the U.S. government, by taking this risk, helped to transform a nation of renters to a nation of owners.

The pursuit of profit by corporations builds effectively on human motivation and has generated far more positive than negative outcomes. But we must also be realistic here. If we take the value of individualism to an extreme, would we also need to accept the behavior of the schoolyard bully? While I deny that such behavior is acceptable, I also do not expect the bully to police himself. Corporations are expected to try to maximize profits, but the government must also set rules for acceptable behavior in corporate competition. Self-regulation generally does not work, and the capitalist system neither expects nor requires us to depend on the good-will of corporations for environmental protection. Only when conserving the environment will help a company maximize its profits or market share will it tend to protect the environment. Otherwise, a system of effective, enforceable rules is needed to govern how companies conduct business. Toxic waste is a natural by-product of an unregulated, technologically oriented capitalist system. If we want our environment to be less lethal, we must have laws and regulations that influence corporate behavior.

The combination of an ignorance of ecology and a glorification of private enterprise permitted corporations to dispose of waste however they pleased, thus creating the dilemma of toxic waste. What values allowed the problem to become a political issue? In some ways it was the dissonance of two sectors of society pursuing the same values, namely, the maintenance and use of private property. One reason toxic waste emerged as a political issue is because the waste was invading suburban homes. Ordinary citizens began to organize against private corporate behavior which they now saw as a threat to their property, health, and security (Kraft and Kraut 1988, 63–65).

This brings us to the third value dimension of the toxic waste problem: the protection of private health and well-being from external threats. Toxic waste is not like a termite invasion; it does not just destroy the frame of your house but can also poison your children or a pregnant woman. Thus, if you are a parent, you not only wish to protect your property but you also seek to protect your children. Here we are operating at a level that may go deeper than values or cultural norms; we are dealing with emotions that are fundamental to human behavior. The drive to protect one's offspring is a biological imperative that appears to be universal to our species. Few policy issues are as intensely political as those that are perceived to threaten our children's health.

In sum, the value dimension of the toxic waste problem is strong and can lead to a clash of deeply held views. The issue is so complicated that key aspects of the problem can be explainable by each facet of the framework developed here.

Toxic Waste as a Political Issue

As I mentioned earlier, toxic waste, in many ways, is custom-made for the U.S. political system, a system designed to maximize the representation of various geographical districts. As a federal system, specific powers are reserved for state governments, and then states delegate certain powers to local jurisdictions (Lowry 1998, 748). Featured in the system are single-member districts and winner-take-all representation. Thus a political party will not benefit from receiving only 20 percent of votes even in every district, for, in all likelihood, the result will be no legislative representatives from that party. Our presidential elections, which are based on the Electoral College, also require that candidates consider geography. As Al Gore can tell you, winning the most popular votes nationwide is not enough; pluralities must be concentrated in specific locations in order to win the all-important state electoral votes. The late Thomas (Tip) O'Neill, long-time Speaker of the House of Representatives, once said that "all politics is local." What he meant is that all *American* politics is local. The absence of proportional representation in our political system discourages extreme positions and favors moderate candidates who can appeal to the broadest possible constituency. Coupled with the geographic bias, environmental political issues often find their way to the political agenda as demands that private property or a family's health be left unharmed. The general issue of ecological well-being or the protection of nature is not raised, as it lacks the political potency of protecting one's family and property.

Toxic waste, unlike climate change or air pollution, is not caused by factors distant or difficult to see; its origin, instead, is both local and visible. The closer you are to its source, in fact, the stronger its impact tends to be. If you are a legislator from a district with toxic waste sites, you are expected to respond to your constituents' demands for government action. These are not ideological issues or ancillary political matters for these representatives. Addressing toxic waste is a matter of serving one's constituency, and elected representatives ignore such issues at their own peril.

When the EPA first proposed the original Superfund legislation, a key part of its presentation on Capitol Hill was a map of the United States by congressional district. Every known waste site over a certain size was represented by little red flags. Needless to say, very few districts were without a few little red flags. Not surprisingly, congressional Representatives paid close attention to this aspect of the presentation (Landy, Roberts, and Thomas 1990, 141).

Local community groups, such as the one led by Lois Gibbs, drove the politics of hazardous waste. Initially all hazardous waste politics was local, but as state and federal governments were called on to shape a response, the form the response took led to a national politics dominated by Washington-based interest groups. Industry organized to influence the shape of the initial Superfund bill and its subsequent revisions. The Chemical Manufacturers Association was a major lobbying force, as were a number of national environmental groups including the Environmental Defense Fund (now Environmental Defense). Lois Gibbs herself moved to Washington and formed a national organization to lobby there and help local communities organize to compel cleanup of toxic sites (Lowry 1998).

The original Superfund bill was the result of a last-minute deal, in late 1980, between then congressman James Florio of New Jersey and the conservative Republican senator Jesse Helms. The bill represented a set of compromises but included substantial funds for cleanup, a tax on "polluters," and strict liability requirements (Hird 1994, 9–10). It was passed in a lame duck Congress in December of that year and was put into place by an executive order signed by President Jimmy Carter on his last day in office. Earlier that fall the bill appeared deadlocked, but pressure from local representatives and homeowners was too great for Congress to resist. Federal legislation would become a model for state-level Superfunds that would provide additional resources and authority for cleanup.

The political force that generated these new programs would become a permanent part of U.S politics in the 1980s, 1990s, and into the twenty-first century. The issue on the political agenda did not focus on the cause of the pollution. In fact, the liability provisions of the statute called for joint, strict, and several liability. In operational terms, this meant that if a private party was found to be responsible for contaminating any part of a site, that party would be held responsible for cleaning up the entire site. It was then the task of that party to find other parties that were also responsible and sue them to recover their share of the cleanup cost (Hird 1994, 17). The government only needed to find one culprit. The direct effect of this bill was to focus the policy agenda on cleaning up the site rather than assessing the cause of the damage, with the result that the volatile local politics of toxic waste centered on cleanup rather than prevention.

Preventing toxic waste requires a more fundamental reexamination of our economic way of life. Consider the regulatory regime involved in the introduction of new drugs. Before a new drug can be marketed, it first has to be tested and proven safe for human consumption. An examination of the drug's primary effects as well as side effects is required. This is called the "precautionary principle." No similar requirements exist for new technologies or products that are not designed to be ingested. As a result, government regulators must play catch-up with the companies making the new substances, and regulation is not permitted until harm is proven. In this respect, we are all the proverbial canary placed in a coal mine to test its toxicity: if a product or technology makes us sick, a remedial response is required; otherwise it is business as usual. Had the policy issue been defined as the need to eliminate the use of toxic substances rather than to clean up mishandled toxic waste, economic interests may have been fundamentally threatened. Similarly, during the early days of the clean air debate in 1970, there was discussion in Congress about banning the internal combustion engine. When the discussion became serious, the automobile companies began to see the merit in setting national ambient air-quality standards.

The basic issue of toxic waste reduction and elimination did reach the political agenda through discussions related to the regulation of hazardous waste prior to the enactment of RCRA in 1976 and the Hazardous and Solid Waste Amendments (HSWA) in 1984. But those discussions had little of the urgency or resources generated by the politics of Superfund. This lack of political investment was not necessarily detrimental to the production of good public policy. Over a long period—nearly a quarter-century—industry learned to pay attention to toxic waste reduction. The changing approach to toxic waste management resulted from several factors. First, increased competition spurred manufacturers to cut costs by reducing waste. New industrial processes such as closed-system engineering made it easier for companies to control their output of waste. Second, de-industrialization contributed to an overall reduction in toxic waste, as more and more industrial businesses closed or moved waste-producing activities overseas. An entirely new field of engineering, called industrial ecology, was created to develop waste-free production processes. In contrast, the Superfund program had to learn how to clean up toxic waste sites in plain public view and found all its mistakes reported on the front pages of the nation's newspapers.

No discussion of toxic waste politics would be complete, however, without taking a look at the NIMBY syndrome, the community-based politics of refusal. The philosophy of "not in my backyard" opposes any kind of development in a given locality and includes resistance to building everything from a sewage treatment plant or an incinerator to a new high school. The politics of refusal is powerful and is one explanation for the intense focus on cleanup in the politics of the Superfund program. The idea that supports that focus, like what motivates NIMBY, is the desire to maintain the status quo or, in the case of cleanup politics, to return to what we thought was the status quo (Mazmanian and Morell 1990, 234–235). Preventing the production of toxics by controlling what a private firm produces is a difficult argument to make in the political culture of the United States. On the other hand, mitigating the damage to private property caused by the production of a toxic is a far stronger argument as it asserts the rights of private property.

The growth of America's suburbs, and now exurbs, is based on a desire to escape the perceived negative impacts of industrial development. The problem is, of course, that industrial development has simply followed people from urban areas to places that once were rural. One often hears a remark like the following: "When I moved here there used to be a forest where we hiked but now it's a mall." The trend is always toward more development and less "natural" space, but as people see the natural amenities they value disappearing, they respond by arguing passionately against additional development.

When a formerly pristine area has been polluted by toxic waste, we see a powerful desire to clean up the mess and restore the area to health. The wish to return such an area to its prior state has a basis in the protection of property rights, but it is also motivated by a sense of social, cultural, and emotional attachment to a way of life that, in the popular imagination, is associated with places relatively untouched by humans. These associations add a level of intensity to the politics of cleanup that elected leaders have found difficult to resist.

The political dimension of the toxic waste problem helped to create a new definition of environmental politics in the United States. It raised the stakes and visibility of environmental politics by adding a health dimension and vastly intensifying the heat of local environmental politics. The initial objective of the Superfund program was to remove people from harm's way and ensure that they were not exposed to toxic chemicals. Politics expanded this objective to include something we did not know how to do, namely, site cleanup (Mazmanian and Morell 1988,82–84).

Toxic Waste as an Issue of Science and Technology

The toxic substances buried in toxic waste sites are by-products of the lifestyle we enjoy in the developed world. Toxic, in this sense, refers to materials that cause harm to human or ecological health. The very technology that affords us the mobility, comforts, entertainment, out-of-season produce, and a wide array of other benefits also creates toxic waste.

Just as technology creates toxic substances, so, too, can it reduce their negative effects. Technology also uses these toxic by-products to reduce their potential for doing harm. Medical technology can mitigate the health effects of toxins. A significant problem with hazardous waste is that when it is dumped on the ground or buried, we lose control of its interactions with other chemicals and with natural ecosystems. If the waste stream of one product can be used as a raw material for another product, the end use of the toxic chemical may be beneficial rather than harmful. Today engineers consider waste reduction and closed-system engineering to be a desirable and feasible goal (Hawken, Lovins, and Lovins 1999). In new management systems, such as Total Quality Management, the reduction of waste in time, labor, and materials is a primary tool for lowering production costs and raising the quality of products (Cohen and Brand 1993, xii; Deming 2000). From the beginning of the industrial age until the 1980s waste was considered to be part of the cost of production. It took a generation of educating people about ecology and the threat of liability costs for toxic waste damage to change the engineering paradigm so that the focus is on waste reduction (Blackman 2001, 197–199).

The data in the developed world clearly indicate that over the past two decades we have seen a significant reduction in the amount of toxic waste created and disposed. Some problems with toxic waste inevitably persist, because production processes continue to produce wastes whose effects are not understood. The technological advancements that have historically contributed to the production of toxic byproducts will not go away. We may think that a certain method can safely manage a waste stream only to learn later that we did not understand the dangers that method posed. Although we are far more careful now than in the past about toxic-waste management, we still do not possess perfect knowledge of the technologies we are using (Davies and Mazurek 1998).

Even if we do a better job of eliminating new toxic waste streams, and even though new engineering techniques can slow the creation of new waste sites, we must still address the legacy of the poor waste management practices of the past. At the start of the twenty-first century we discovered that some of the sites on the Superfund priority list were very large and relatively old. Toxic waste becomes more difficult to clean the longer it is in the ground. Although water and gravity can dilute toxicity, they also cause waste to spread over a wider area, complicating cleanup efforts. Cost estimates for cleaning these so-called mega-sites ranged from $50 million to $100 million (U.S. EPA 2001).

While the problems associated with cleanup get more complicated over time, two types of technological fixes are possible. One is to develop methods to detoxify the waste when it gets into water or air or near people. The other is to develop methods of reducing the health effects of exposure. It is true that many of the chemicals in the waste stream are carcinogenic. If medical research develops a cure for cancer, those substances may come to be seen as less toxic than they are viewed today. The development of medical technologies to mitigate the physiological effects of environmental toxins may prove to be an effective means of addressing the problem of hazardous waste. Despite the increased toxicity of our planet over the past fifty years, people in the developed world are leading longer and healthier lives. Thus the health benefits of science, at least for now, are exceeding the costs of its hazardous by-products.

Containment of toxic waste contamination is one of several methods currently used to clean up hazardous waste sites. The technology of containment has developed sufficiently to allow for the redevelopment of many urban "brownfields," the term for old, abandoned, and sometimes contaminated industrial sites. Under legislation passed in 2002 (Public Law 107–118, "Small Business Liability Relief and Brownfields Revitalization Act"), the standards for site cleanup are made more flexible and can differ according to the proposed use of the site. An interesting aspect of the development of cleanup approaches is the degree to which they are based on previous EPA programs, especially the sewage treatment program. Sewage treatment and Superfund were the only two multibillion-dollar programs in the EPA, and the same career civil servant, Michael Cook, ran the sewage treatment program in the 1970s and then ran the Superfund program in the early 1980s. Large-scale engineering construction firms became the prime contractors for both programs. In both cases the technology used involved constructing thick concrete walls and filtration systems.

Some of the people involved in both EPA programs had previously been contractors and staff in Army Corps of Engineer infrastructure projects. The technological base of the Corps of Engineers dam-building program involved moving earth and construction. Through the latter part of the twentieth century the technology of construction improved, and, as engineers gained experience, site cleanup technology also developed. However, the EPA did not emphasize the use of nonconstruction technology such as biotechnology and other potential areas of detoxification. Since the job of the Superfund program was to keep people out of pathways of exposure, the EPA's approach was sensible and prudent, but it also provided a specific direction to the program's approach to technology.

The reuse of brownfields and the increasing political acceptability of risk assessment as a basis for decisions regarding land use may serve to reduce some of the "heat" from the politics of toxic waste cleanup (Bartsch and Collaton 1997, 31). If people become accustomed to the idea that the risk of buried toxic waste might be acceptable under particular conditions and certain uses, the issue may lose the black-and-white clarity we saw at Love Canal. This political process may already be under way, made possible by improvements in waste containment technology.

Toxic waste creation and cleanup are clearly issues with a strong basis in science and technology. The problem is caused in part by a lack of understanding of technological by-products, and we count on technological solutions to fix it. The past damage of inadequate waste management will probably never be amenable to single or simple technological solutions. The waste materials and the environment they are released into vary widely. Each toxic waste cleanup initiative must be customized to its particular mix of pollutants and environments.

When Superfund was enacted in 1980, the EPA knew almost nothing about toxic waste cleanup. A quarter-century later we have learned a great deal about the dilemma, but we still lack an adequate command of the science and engineering of toxic waste cleanup. The policy design and management issues involved are exacerbated by the cost of cleanup and lack of scientific certainty about the problem and its potential solutions.

Superfund and Toxic Waste Cleanup as an Issue of Policy Design

The program design of hazardous waste policy includes both a regulatory dimension and a direct-action emergency response function resembling that of a fire department. The regulatory side involves regulations in HSWA and RCRA on the transport, storage, and disposal of hazardous waste. Superfund also includes regulations related to cleanup. These rules collectively govern when, what type, and the degree of cleanup that is required of a private party for a particular site.

Regulations related to the transport, storage, and disposal of waste are traditional command-and-control regulations. Private companies must receive permits from the EPA to handle and dispose of waste. Inspections are conducted to determine adherence to the permits. If violations are detected, the EPA and the courts seek to impose penalties. In most cases, administration of the regulatory program has been delegated to state governments. This type of system has a number of problems. First, the expense of administration and inspection is high and must be borne by the government. Second, other than fear of getting caught, there is little disincentive for "midnight dumping" or other forms of illegal waste transport, storage, or disposal. This aspect of the program certainly calls out for some creative policy design work.

A mechanism that contains a profit incentive for private parties to properly dispose of waste is needed. Examples might include a tax benefit for demonstrated waste reductions or a requirement to insure waste for safe and verified receipt at an approved waste disposal facility.

Economists often think that the need for government regulation arises as a result of market failure (Nakamura and Church 2003, 15). In this framework, consumers of the environment (i.e., people) have a theoretical willingness to pay for maintaining their environment in a nonlethal condition. When the production process includes actions taken to reduce negative environmental impacts of either the product or production by-products, the price of environmental protection is borne by producers who recoup the cost by raising the price of their product. Consumers who are highly willing to pay for environmental protection will respond, theoretically, by paying more to purchase a product less harmful to the environment. According to this model, a market system, created through the provision of incentives or the imposition of disincentives, offers an effective and efficient means of achieving the intended policy goal of environmental protection. Command-and-control regulation, on the other hand, is seen as an expensive and inefficient way to implement public policy. I should note that one part of Superfund, its emergency response provision, is not regulatory but is a government direct police function. Thus Superfund's policy design includes nonregulatory and regulatory elements.

While Superfund's regulatory design is quite traditional, I do not want to fall into the equally traditional academic practice of criticizing the government for this design. Early on, we did not have the experience or information needed to develop a more sophisticated or elaborate policy design. The toxicity of some of the chemicals in the soup we were trying to bring under control was high and not well understood. One great advantage of command and control is that it establishes a principle of law and allows an initially unambiguous definition of correct and incorrect behavior. I would argue that, for the most part, command and control is a necessary first step in a new area of regulation. The extended amount of time needed to put RCRA in place is an indication of the information and capacity that had to be developed and were not in place in the late 1970s. Moreover, some market failures cannot be corrected, or, if they can, the costs of correction are higher than those of using other regulatory devices.

Long-term toxic waste cleanup is an example of the very complex policy design in the original Superfund bill. Three categories are used to define the status of a site. The first distinction is between emergency response (or threat removal) and long-term remedial response (the extended cleanup of a waste site). The second is the distinction between spills from transportation and leaks from stationary sites. Third is the differentiation between sites and spills that have financially viable responsible parties to pay for cleanup and abandoned sites for which no owner can be assigned responsibility (U.S. EPA 2003b).

The Superfund program had to be designed to accommodate each of these types of cleanup situations. More recently an additional concern has arisen for large-scale, complex "mega sites." As noted earlier, existing capacity in sewage treatment and engineering was used to develop methods of remedial response—or site cleanup. Similarly, the emergency response program grew out of the capacities of EPA's oil spill response program.

Emergency Response under Superfund

The first head of EPA's oil spill program, Kenneth Biglane, came to Washington from Louisiana where, as early as the 1950s, he worked on oil pollution control. When Superfund was created, Michael Cook recruited him to run the Hazardous Response Support Division (U.S. EPA 1999, 3). The part of Superfund that called for fire department—like response services was implemented through this facet of the organization. In the program's early years, Biglane assembled regional Emergency Response Teams and a network of contractors to clean spills and stabilize sites—measures developed to realize Superfund's mission to ensure that people were not exposed to toxic waste. Also early on, while the long-term cleanup program struggled with a range of complex issues, such as when a site is clean enough to reuse, this group succeeded in executing thousands of removal actions as well as ensuring that most of the toxic waste in the United States did not leak into groundwater and people's basements. Sometimes this involved activities as simple as moving tipped over barrels containing toxic waste and placing them upright on a strip of asphalt. At other times fences were built around waste sites to keep children from playing there. Also, there were coordinated responses to oil and chemical spills. Much of these Superfund activities were performed out of the view of the media, which focused instead on long-term cleanup (Hird 1994, 29).

By bringing an emergency response function into the program, the EPA was able to provide a response to urgent public demands even though the technology of toxic waste cleanup was quite primitive. We might not know how to clean up a toxic waste site, but we could measure the chemical composition of land and do some of the simple housekeeping tasks needed to contain the damage. We also knew how to reduce the speed of waste migration off-site and how to move people from their homes when threatened with contamination. One of the great paradoxes of the Superfund program is that although the program failed in its cleanup endeavors, it succeeded in threat removal. By borrowing organizational models and standard operating procedures from oil spill cleanups in the Gulf of Mexico and from urban emergency medical services and fire departments, the policy design was able to substantially reduce the dangers of human exposure to toxic waste in the United States.

Long-Term Cleanup, or Remedial Response, under Superfund

Site cleanup, or remedial action, is the most complex element of hazardous waste program design. When a toxic site is uncovered, an early step is the attempt to assign responsibility for the damages. This can be done by investigating the history, ownership, and uses of a given site. Superfund's liability provisions have given the government the ability to compel private cleanup and recover the costs of government-conducted cleanup from private companies. The ability to recover the costs of cleanup, however, requires the government to establish and maintain an incredibly detailed set of financial records for each aspect of site cleanup, which adds to the time and cost.

In order for Superfund to take action on a site, the site must be assessed by the EPA and receive a high rating on EPA's Hazard Ranking System. If the site is considered sufficiently serious to merit action, it is either placed on the NPL, referred to another environmental cleanup program, or named a Superfund Alternative Site. This last category is for sites with cooperative responsible parties (U.S. EPA 2004).

The remedial program was designed with several objectives in mind. The first goal was to clean up or at least stabilize the worst sites. The second was to try to maximize the impact of public funds by leveraging as much private cleanup as possible. A third aim was to ensure that the cleanup was effective in eliminating the spread of contamination and, where possible, permitting the reuse of the land. A problem, however, is that the program's design is characterized by a great deal of technical, environmental, and legal uncertainty. Consider, for example, the following key questions, to name just a few:

- What do we mean by the "worst sites"? It took years to develop a workable hazard ranking system.
- How clean is clean? Or, in other words, when is a waste site cleanup completed? And are there different cleanup standards in different environments and for different end-uses of the land?
- Who pays for the cleanup? And how are costs to be apportioned to different private parties responsible for the damage?
- Who is liable for damages to property and health resulting from a site's contamination? And what is a reasonable level of compensation?

Superfund is often derided as the Environmental Lawyers Full Employment Act. The amount of litigation associated with toxic waste cleanup has been extensive and a source of frustration to the private firms footing the bill, and to environmental and community advocates who just want to see the site cleaned up.

The idea that this litigation or the program's complexity is a design flaw should be resisted. When Superfund was enacted in 1980, very little was known about hazardous waste. To our collective horror, we had suddenly discovered that there were tens of thousands of hazardous waste sites in the United States. Not only had private companies deposited toxic waste all over the landscape, the military and the federal government had probably dumped even more (Barnett 1994, 21). To provide perspective on how little we knew, here is a brief story from my time working at the EPA. In late 1980 my boss, Marc Tipermas, the director of Superfund's policy office (the Office of Analysis and Program Development), assigned me the task of developing a policy statement on federal facilities. I asked him what he meant by "federal facilities," and he responded by saying that he suspected the federal government had toxic waste sites of its own and that eventually we would be asked about EPA's policy toward the waste dumps created by our colleagues in the federal government. He thought it might be a political problem that we were asking others to clean up their waste when our own house was not in order. In retrospect, I see that this was a farsighted observation. At the time, however, I was somewhat miffed, thinking it was an unimportant assignment. I could not have been more wrong. During the 1990s, $33 billion was appropriated to the Departments of Energy, the Interior, Defense, and Agriculture for hazardous waste cleanup. A multi-agency task force estimated that the total costs for cleanup of federal facilities were between $234 billion and $300 billion over a seventy-five-year period (U.S. Government Accounting Office [GAO] 1999, 14–15).

I was not alone in my ignorance. The EPA, in general, had very little knowledge or understanding of the problem of hazardous waste (Barnett 1994, 10). In cases such as the early Superfund program, a key objective of toxic waste management program design was to facilitate organizational, technical, political, and social learning. Before a problem can be solved, it needs to be understood. The more complex and deep-rooted the problem, the longer the learning process. In the case of toxic waste, program design has facilitated an enormous amount of learning.

The policy design issue is complicated by the need to clean sites while preventing new sites from becoming contaminated. Further complicating the issue is that the longer one waits to clean up a site, the more damage is done to the environment and the more expensive the cleanup becomes. The second EPA assistant administrator to whom Mike Cook reported on Superfund was Eckhardt C. (Chris) Beck. Beck had an expression for his policy preference on this aspect of the program. He said that Superfund should follow a practice of "shovels first and lawyers later." Let's first reduce the risk and then worry about who pays for it, first put out the fire out then figure out whose fault it was. Indeed, that was the original idea behind Superfund. This was not the time to allow dollars to drive the program. Despite the mountain of red tape and complexity that followed, the design was successful in one important respect: the nation's waste sites were largely contained and the health of the general public was protected. The mix of emergency response, private- and government-funded cleanup, command-and-control regulation, and fear of private liability judgments improved the way hazardous waste was managed and allowed the long process of cleaning and containing America's waste problem to get started.

Toxic Waste Cleanup as a Management Issue

While other parts of this chapter combined the waste regulation aspects of RCRA/HSWA with the cleanup and response dimensions of Superfund, here the focus is specifically on the organizational capacity to respond to toxic waste cleanup. Recall that cleanup capacity must include three dimensions: emergency response, remedial response, and enforcement and cost recovery.

Emergency response was the part of Superfund that had the clearest and most appropriate model to borrow and build on. Ken Biglane and his colleagues had been in the business of responding to oil spills for several decades, and focusing their attention on responding to toxic waste emergencies was relatively simple. Greater care was required in handling highly toxic materials, but spill response staff was familiar with working in dangerous situations. As noted, this group thought and acted like emergency workers. The first wireless phones and fax machines I ever saw were in the office of the EPA's Emergency Response Division back in 1980. This group had a different approach than the rest of the career bureaucrats at the EPA. They developed emergency response teams in a number of key EPA regions as well as national contracts that allowed them to provide rapid response capability during environmental emergencies. They could quickly assess a site for the danger posed, and they were able to put together a short-term response that would eliminate a hazard until a long-term response could be organized.

In Superfund's early years, while the rest of the program slogged through the morass of building long-term response capacity, this part of the program conducted thousands of threat removal actions. The sole function of these teams was to stabilize a site; they were not responsible for controversial tasks such as assessing the level of toxicity or assigning responsibility for cleanup. Their job was simple, and no one doubted its necessity: get to the spill or the site and quickly figure out what is wrong. Fix it if possible, or at least keep it from getting worse, and then get out. Emergency response did not require the EPA to develop new management or organizational capacity. It knew how to do this, and it had staff with experience to manage the work.

Remedial response was another story. No one had ever cleaned up and fully restored a waste site before. The closest analogies were dam construction and other large-scale, earth moving activities by the Corps of Engineers and the EPA-funded construction of water filtration and sewage treatment plants. We did not know how to characterize the degree of contamination at a waste site. Several large-scale contracts were made with environmental engineering firms to conduct field investigations at waste sites. Contracts were also developed eventually to design and construct remedial actions to clean these sites. Unlike water supply and sewage treatment, where each plant was more or less the same (with only slight variations), each waste cleanup had to be custom designed. The mix of chemicals, the local ecology, and the pathways of exposure to humans varied at each site. A sewage treatment plant, on the other hand, simply took waste from sewers and filtered it before releasing it into a body of water. Engineering standards were developed over the years for each type of plant situation. In the case of toxic waste, the EPA needed to develop the capacity to assess sites for long-term damage, develop and analyze cleanup design strategies and facilities, and then manage the construction of these facilities. While patterns eventually emerged, and we learned a great deal about different types of sites and potential solutions, in the early 1980s we knew very little. One of the early management problems faced by the Superfund program was the absence of standard operating procedures (SOPs)—preformed responses to different problems at different waste sites. One of the program's greatest accomplishments during its first two decades was the development of SOPs.

The efforts of the Superfund program in the 1980s were largely devoted to understanding the various dimensions of the toxic waste problem and to developing the capacity needed to address it. We did not know how to clean up a site. Nor was it clear that we ever would or even if it was a cost-effective use of scarce environmental protection dollars. Over time we began to learn how to do this work, and we began to discover how complicated and expensive some cleanups could be.

I would characterize toxic waste cleanup as a problem that had a strong management dimension at the outset. In the 1980s we did not have the organizational capacity in place to do the work needed. Today, a quarter-century later, the management dimension to the problem still remains, but it is of declining significance. One thing we have learned since the start of the Superfund program is how very complicated the problem is. We can manage the available technology and contain the waste. We can also use filtration processes, even soil removal and disposal, to detoxify or at least reduce contamination. But site remediation is expensive and complex, so although we have increased our capacity to do the work, there will always be management issues.

Summary of the Multiple Dimensions of Toxic Waste Cleanup

New York City's garbage and America's underground tanks are relatively simple problems compared to toxic waste. When a tank is leaking, we tend to know what was in it. The city's garbage is voluminous, but it, too, is fairly predictable. Toxic waste, on the other hand, introduces a number of different and unpredictable variables. It is not always clear what was buried decades ago in a toxic waste site. Hazardous waste, overall, is a tougher and more complex problem. While we know how to make a tank that doesn't leak and how to safely dispose of household waste, we are still learning how to clean up toxic waste sites. In that respect the issue of toxic waste cleanup is largely a problem of science and technology. To the extent that it is a management problem, it is one of managing a new and unproven set of technologies. If the technology were more familiar, we would have a better idea of the organizational capacity needed to deploy the technology.

In the case of toxic waste cleanup, the political and value dimensions are intertwined. The politics of cleanup can be intense, because the value of defending one's property and protecting one's family are deeply held in this culture. Siting a waste treatment, transfer, or disposal facility in New York City is an inherently hot political issue, but so, too, is siting a school or a big-box discount store. When the issue involves the potential for being poisoned, the politics is even more heated. When the politics of solid waste in New York City becomes really intense, it is often about the perceived toxic elements of handling household waste. In the case of toxic waste, the waste is always poisonous. Issues involving perceived threats to public health arouse high levels of emotional intensity. It is this intensity that shapes how these issues are defined when they reach the policy agenda. Even though Superfund's removal actions accomplished an enormous amount of good by reducing the risk of toxic waste during the program's first two decades, the failure to clean up sites through remedial action was all that mattered politically. Keeping poison away from people was not enough. We had to eliminate the poison. The emphasis on long-term cleanup and land reuse was a direct result of this politically driven definition of the problem.

The problem also retains a strong basis in science and technology, as the toxics are created as a by-product of chemicals used to produce other goods. If and when science develops more effective methods for detoxifying land, the political potency of this issue may be reduced dramatically. The management of new cleanup techniques, however, will remain a challenge. As new cleanup methods are developed, staff will need to be trained to use these methods and organizations will need to develop the capacity to recruit, train, and deploy staff to do the new work.

Conclusions

We have now examined a local issue (New York City's garbage), a "simple" national issue (underground tank leaks), and a highly complex national issue (toxic waste management). It seems clear that a thorough understanding of the toxic waste issue requires that we apply more elements of the framework introduced in chapter 2 than we do for the other issues. While New York City's solid waste is essentially a political issue, and underground tanks predominantly a management issue (with a strong strain of science and technology), toxic waste is fully multidimensional. No single element of the framework dominates. A better regulatory design might reduce the number of sites created in the future, but a strong government direct action and enforcement effort is needed to clean up existing sites. Private voluntary cleanup is unlikely, unless driven by fear of liability from toxics migrating off-site.

Still, it is hard to argue that the design of the convoluted and frequently delayed site cleanup program could not be improved. Given the substantial political, managerial, and technological challenges faced by the Superfund program, it is safe to conclude that no single element dominates. The use of the framework helps us to develop an appreciation for the difficulty and likely long-term persistence of this environmental policy issue.

Chapter 6

Have We Made the Planet Warmer, and If We Have, How Can We Stop?

The Nature of Global Climate Change

The Earth's climate is an extremely complex system, making it difficult to identify trends and their causes. In the last three decades scientists have become increasingly certain that global temperatures are rising. Temperature records and other data reveal, however, that the Earth's temperature has always fluctuated. Separating natural fluctuations from anthropogenic or human-induced change is a major challenge faced by scientists working to interpret recent changes in global average temperatures. The impact of human activities on climate has long been a subject of study. In 1970 Helmut Landsberg, one of the first scientists to identify and quantify such changes, published an article on the "heat island effect," a phenomenon of urban areas retaining more heat than less urbanized areas (Landsberg 1970, 1270). According to the American Institute of Physics (AIP), Center for the History of Physics:

> It had long been recognized that the central parts of cities were distinctly warmer than the surrounding countryside. In urban areas the absorption of solar energy by smog, black roads and roofs, along with direct outpouring of heat from furnaces and other energy sources, created a "heat island" effect, the most striking of all human modifications of local climate. It could be snowing in the suburbs and raining downtown. . . . Some pushed ahead to suggest that as

> human civilization used ever more energy, in a century or so the direct output of heat could be great enough to disturb the entire global climate. (AIP 2003)

In the 1970s scientists were uncertain whether the planet was getting colder or warmer.

In 1975 Wallace Broeker of Columbia University's Lamont Doherty Earth Observatory addressed this issue and posited that some of the cooling tendencies in the Earth's climate cycle were masking heat effects caused by carbon dioxide (CO_2) (Broeker 1975, 460–463):

> He [Broeker] suspected that there was indeed a natural cycle responsible for the cooling in recent decades, perhaps originating in cyclical changes on the Sun. If so, it was only temporarily canceling the greenhouse warming. Within a few decades that would climb past any natural cycle. "Are we on the brink of a pronounced global warming?" he asked. (AIP 2003)

Rising global temperatures have been attributed largely to an atmospheric process known as the "greenhouse effect." Normally molecules of certain atmospheric gases trap heat like the window panes of a greenhouse. This natural greenhouse effect is responsible for keeping the Earth's temperature warm enough to support life as we know it on this planet. Human activities have introduced massive quantities of greenhouse gases, particularly CO_2, into the atmosphere, increasing its heat-trapping capacity. The principal goal of global climate policy is to address the climate impacts caused by increasing concentrations of greenhouse gases.

In 1981 a team led by NASA and Columbia University scientist James Hansen predicted that the dangerously heightened levels of CO_2 would cause global warming by the end of the twentieth century (Hansen et al. 1981, 957–966). Hansen's team observed that,

> Any greenhouse warming had been masked by chance fluctuations in solar activity, pulses of volcanic aerosols, and increased haze from pollution. Furthermore, as a few scientists pointed out, the upper layer of the oceans must have been absorbing heat. These effects could only delay atmospheric warming by a few decades, however. Hansen's group boldly predicted that considering how fast CO_2 was

> accumulating, by the end of the 20th century "carbon dioxide warming should emerge from the noise level of natural climatic variability." Around the same time, a few other scientists using somewhat different calculations came to the same conclusion—the warming would show itself clearly sometime around 2000. (AIP 2003)

Although there is disagreement about what should be done about global warming, the U.S. government and the EPA accept the fact of rising global temperatures. According to the EPA's website:

> The Earth's surface temperature has risen by about 1 degree Fahrenheit in the past century, with accelerated warming during the past two decades. There is new and stronger evidence that most of the warming over the last 50 years is attributable to human activities. Human activities have altered the chemical composition of the atmosphere through the buildup of greenhouse gases—primarily carbon dioxide, methane, and nitrous oxide. The heat-trapping property of these gases is undisputed although uncertainties exist about exactly how Earth's climate responds to them. (U.S. EPA 2000)

The EPA website in 2004 included a major section on climate "uncertainties," and yet the White House believed that the situation was certain enough to adopt a policy to control greenhouse gas emissions. The policy of the Bush White House in 2002 was intended to cut "greenhouse gas intensity" by 18 percent from 2002 to 2012. Greenhouse gas intensity is the ratio of greenhouse gas emissions to economic output. The goal of the Bush administration was to lower the rate of emissions in the United States from an estimated 183 metric tons per million dollars of GDP in 2002 to 151 metric tons per million dollars of GDP in 2012 (White House 2002). Under this policy, greenhouse gas emission reductions or increases are a function of the rate of GDP growth. While it is still unclear whether this policy will actually be implemented or whether it will have an impact on the problem, at least it was an indication that the U.S. administration considered climate change a problem worthy of government action.

Most scientists believe that the current warming trend is at least partly the result of human activities, particularly fossil fuel combustion. The Intergovernmental Panel on Climate Change (IPCC) has stated that the warming of the last fifty years cannot be explained by natural causes alone (IPCC 2001). Global warming is primarily caused by changes in land use and the increased utilization of fossil fuels by the world's more than six billion people. The United States alone emits more than 25 percent of the world's greenhouse gases, making it the most significant contributor to the problem of anthropogenic climate change (IPCC 2001). Just as toxic waste cleanup was an issue custom-made for the U.S. political system, climate change is the quintessential international issue. It cannot be addressed by one nation alone. As scientific uncertainty has lessened, climate change has emerged as an important item on the international institutional agenda, and international discussions are resulting in policy development.

Attempts to initiate international governance of climate change began in earnest when, in 1988, the UN Environment Programme (UNEP) and the World Meteorological Organization (WMO) established an intergovernmental working group to prepare for treaty negotiations (Information Unit on Climate Change [IUCC] 1993). In 1990 the UN General Assembly established the Intergovernmental Negotiating Committee for a Framework Convention on Climate Change (INC/FCCC). Diplomats from more than 150 nations met six times between February 1991 and May 1992 and, in May 1992, adopted the United Nations Framework Convention on Climate Change (UNFCCC). In June 1992, at the UN Rio Earth Summit, the agreement, which set the general parameters for an international policy on climate change, was signed by 155 nations. To stabilize greenhouse gases, the framework established six guiding principles and three areas of obligation. The principles were the following:

1. *Common but differentiated responsibilities.* Countries would be asked to contribute to emissions reductions according to their level of development.
2. *Equity of requirements,* particularly between poor and rich countries, would be critical to achieving an international agreement on greenhouse gas emissions and reductions. Interpretations of what constitutes equity vary widely.
3. *Precautionary principle.* It is better to be safe than sorry. Complete scientific certainty is not needed to act on climate change.
4. *Cost-effectiveness.* The policies adopted to reduce emissions should bring the most benefit for the smallest possible cost.
5. *Sustainable development.* Each nation has the right to pursue sustainable development.

6. *An open international economic system.* This is a reiteration of the principle of free trade.

The three basic obligations imposed by the framework, principally on developed nations, were these:

1. *A gradual return to 1990 levels of greenhouse gas emissions.*
2. *Provision of financial resources and technology to developing countries in order to promote sustainable development.*
3. *Provision of data on emissions and mitigation efforts.*

In December 1997 the third meeting on the Framework Convention on Climate Change was held in Kyoto, Japan, where the parties adopted the Kyoto Protocol, an agreement specifying emission reduction targets for different countries. The targets were designed to allow for international progress toward the goals of the framework. For countries ratifying the Protocol, emissions targets would become binding. Greenhouse gas emissions would be reduced by an average of 5.2 percent below 1990 levels by 2012 (UNFCCC 2002). Under the Clinton administration, the United States signed the Kyoto Protocol in 1997, but the Protocol was never introduced to the U.S. Senate for ratification. Instead, by a vote of 95 to 0, the Senate passed the Byrd-Hagel Amendment in July 1997, which expressly requested that the United States not enter into any treaty requiring reductions that could prove damaging to the economy or would not hold developing nations to the same commitment schedule as developed nations. Despite this setback, hopes for Kyoto remained. But with the election of George W. Bush these hopes were dashed, when the president, shortly after taking office, announced his intention to abandon the treaty entirely. Notwithstanding the U.S. stance on Kyoto, Russia ratified the treaty in November 2004, allowing it to take effect in February 2005.

As noted above, the Bush administration's policy on climate change in 2002 called for a reduction of greenhouse emissions relative to GDP growth. The program did not mandate any specific reductions but proposed to employ the following mix of programs to reduce emissions:

- "**Substantially Improve the Emission Reduction Registry.** The president directed the secretary of energy, Spencer Abraham, to propose improvements to the current voluntary emission reduction registration program under section 1605(b) of the 1992 Energy Policy Act.

• **Protect and Provide Transferable Credits for Emissions Reduction.** The president directed the secretary of energy to recommend reforms to ensure that businesses and individuals that register reductions are not penalized under a future climate policy, and to give transferable credits to companies that can show real emissions reductions.

• **Review Progress Toward Goal and Take Additional Action if Necessary.** If, in 2012, we find that we are not on track toward meeting our goal, and sound science justifies further policy action, the United States will respond with additional measures that may include a broad, market-based program as well as additional incentives and voluntary measures designed to accelerate technology development and deployment.

• **Increase Funding for America's Commitment to Climate Change.** The president's FY '03 budget proposed $4.5 billion in total climate spending—an increase of $700 million. The budget proposal:

1. Dedicated $1.7 billion to fund basic scientific research on climate change and $1.3 billion to fund research on advanced energy and sequestration technologies [which are technologies for removing gases from emissions and storing them].
2. Included $80 million in new funding dedicated to implementation of the Climate Change Research Initiative (CCRI) and the National Climate Change Technology Initiative (NCCTI). Funding was designated to be used to address major gaps in current understanding of the natural carbon cycle and the role of black soot emissions in climate change. It will also be used to promote the development of the most promising "breakthrough" technologies for clean energy generation and carbon sequestration.

• **Implement a Comprehensive Range of New and Expanded Domestic Policies**, including

1. *Tax incentives for renewable energy, cogeneration, and new technology.* The president's FY '03 budget proposed $555 million in clean energy tax incentives, as the first part of a $4.6 billion commitment over the next five years ($7.1 billion over the next ten years). These tax credits will spur investments in renewable energy (solar, wind, and biomass), hybrid and fuel cell vehicles, cogeneration, and landfill gas conversion . . .

2. *Business challenges.* The president has challenged American businesses to make specific commitments to improving the greenhouse gas intensity of their operations and to reduce emissions . . .
3. *Transportation programs.* The Administration is promoting the development of fuel-efficient motor vehicles and trucks, researching options for producing cleaner fuels, and implementing programs to improve energy efficiency . . .
4. *Carbon sequestration.* The president's FY '03 budget requested over $3 billion—a $1 billion increase above the baseline—as the first part of a ten-year (2002–2011) commitment to implement and improve the conservation title of the Farm Bill, which will significantly enhance the natural storage of carbon . . ." (White House 2002)

This collection of policies and programs seemed more like a public relations initiative designed to placate critics of U.S. climate policy than a coherent effort to reduce greenhouse emissions. The policy remained unchanged at the start of President Bush's second term as he continued to resist the type of policy design that would encourage industry to reduce CO_2 emissions or develop means of sequestering CO_2. The Bush administration clearly saw a trade-off between reducing global warming and economic growth. The administration's refusal to regulate was based on the lobbying power of the energy industry, making the argument that reduction of fossil fuel use would reduce economic growth (Moroney 1998). Rhetoric aside, the intention of the U.S. policy on climate change from 2001 through 2005 was to avoid requiring American businesses to curb emissions of greenhouse gases. The hope was that research on new technologies would generate alternatives that would allow the United States to continue its pattern of consumption with less impact on the environment.

Climate change obviously is not just a problem in the United States. The European Union has supported the Kyoto Protocol and is working to meet its targets. In a speech to the Second Brussels Climate Change Conference on May 11, 2004, Catherine Day, Director General for Environment of the European Commission observed that,

> EU policy on climate change is based on two main principles: Firstly, we must employ market forces in the search for the lowest cost

opportunities to reduce greenhouse gases. Secondly, we must engage all sectors of the economy in meeting national reduction targets . . .

The European Climate Change Programme is based on these two principles and our estimates demonstrate that achieving the EU's Kyoto target will be affordable for the EU economy. Overall the ECCP [European Climate Change Programme] identified measures with a potential to meet approximately twice the reduction we need to achieve under the Kyoto Protocol.

She concluded by noting that,

Climate change has to be addressed globally. There is a general recognition that the Kyoto Protocol targets are only a first, small step to addressing the long-term climate challenge. . . . Much deeper emission cuts will be needed. The only way forward will be through the involvement of all developed countries, and the provision of sufficient incentives to allow developing countries to engage in the fight against climate change without compromising the fight against poverty. (Day 2004)

The problem of climate change requires a high degree of international cooperation. By mid-2004 a clear international consensus emerged on the need to reduce CO_2 emissions to slow the process of human-induced climate change (Betsill 2004, 118). The clearest expression of this consensus in the United States was the bi-partisan Climate Stewardship Act proposal from Senators John McCain and Joseph Lieberman. Although the proposal was defeated in late 2003, it continued to set the policy agenda in Washington on the issue of climate change throughout 2004 and 2005. The legislation proposed to set limits on releases of CO_2 and five other greenhouse gases, and it also proposed the creation of a market where firms or industries that could not meet reduction goals could purchase "credits" from those that managed to exceed them. The bill called for a reduction of greenhouse gas emissions to 1990 levels by the year 2016. While less ambitious than the Kyoto Protocol, it would commit the United States to a specific, mandated timetable.

A fundamental policy problem in international relations is that it is difficult to compel sovereign nation-states to comply with rules established at the international level. Still, it is not impossible. International treaties constrain the behavior of powerful sovereign states. A culture of global norms can create effective taboos, such as the one that has discouraged the use of weapons of mass destruction. The potential for catastrophic failure can make the development of the taboo more likely but also more necessary. The reduction of greenhouse gases will probably require large-scale, gradual change. This can and should be monitored and encouraged internationally. But if the largest emitter of CO_2 refuses to participate in these reduction efforts, the rule's legitimacy is undermined. The international nature of this problem creates governance issues that are inherently difficult to address.

Global Climate Change as an Issue of Values

One cause of climate change, as we saw in all the environmental problems discussed so far, is our lifestyle choices, which are determined by our values. Behaviors based on these values have resulted in population growth and extensive use of fossil fuels. The philosophy of living for the moment that pervades our culture limits our interest as a society in addressing climate change. The true dangers of global warming will occur in the future. Because the effects cannot be predicted precisely, they are removed from current reality. The public responds far more quickly to environmental problems that immediately affect their personal lives. Air and water pollution and the hazard of toxic substances provoke a faster and more active response than climate change because the impacts of the former are immediate.

The effects of global climate change are complex and difficult to see. Many of them will not become apparent for several decades. Social learning is required if climate change is to achieve status on the systemic and institutional policy agenda. We need to imagine that we ourselves are living in the future and experiencing the negative impact of global warming. Analysts who examine the solvency of the social security system face the same challenge, for most of the people who will be harmed if the system goes broke have more immediate financial worries to focus on and so the system's solvency does not get on the policy agenda.

Another value dimension to the issue stems from the question of equity. Do people in the developed world have an unlimited right to burn fossil fuels to maintain their lifestyles while denying that same right to people in developing countries who are aspiring to a more consumptive lifestyle? A second equity concern is the greater vulnerability of poor people to the effects of climate change. Very often a wealthy nation can defend its settlements and food supplies against the negative impact of an extreme climate event. For example, the effects of droughts can be mitigated by irrigation, and the damage from floods can be fixed if massive amounts of capital are available to pay the costs of reconstruction. Poorer nations, on the other hand, lack the resources required for these kinds of responses. In his book *American Heat: Ethical Problems with the United States' Response to Global Warming,* Donald A. Brown (2002) takes the position that, because global warming has had and will continue to have a disproportionately large negative impact on poorer countries, reductions in emissions are a moral imperative. To Brown, the question that remains is how much reduction is needed, and, in his view, "no matter which ethical rule is followed on deciding on an atmospheric stabilization goal, the status quo on global warming emissions is ethically reprehensible" (232).

I do not agree that the ethical dimensions of climate change are as stark as Brown believes, but that ethics is one facet of the problem is clear. Because the science of climate change is uncertain, and because much of the impact will occur only in the future, I find the ethical dimension of the issue somewhat ambiguous. The key moral issue posed is that poor nations and poor people will probably be disproportionately affected by the climate change. Such change is also likely to worsen the environmental, social, and economic problems that lead to extreme poverty in the first place. Greenhouse emissions are clearly harmful, and we have an ethical obligation to reduce the threat of global warming for the most vulnerable among us, as well as for future generations. The political potency of this issue is rooted in its ethical dimensions.

Global Climate Change as a Political Issue

The industries that stand to lose the most from having limits set on greenhouse gas emissions are those that produce oil and energy (Levy and Newell 2004, 194). The Bush administration has evinced extremely close ties to both, particularly the oil industry. A powerful force motivating the administration to question the science of global climate change was the financial and political power of the American oil industry. It was not surprising, of course, that President Bush, the former governor of a major oil state with a large base of campaign contributors in the oil industry, was on the side of the oil industry with regard to reducing greenhouse emissions despite campaign promises to the contrary in 2000. The fundamental logic of this position is that reducing greenhouse emissions will cost money that could otherwise be spent by consumers on goods or services or by the government on services or infrastructure, which would earn elected leaders political credit from constituents. The other side of the political equation is that enforcing emission reduction policies would gain elected leaders credit from environmentalists.

Climate change is more a science and technology issue than a political one. No one is in favor of global warming, but those opposed to strong measures to reduce CO_2 emissions tend not to believe that the cost of such measures outweighs the benefits. Although certain businesses will be disadvantaged by the costs, overall it appears that higher environmental standards have indirectly advanced the modernization of American industry, creating wealth by creating cleaner environments. The operation of sewage treatment plants, for example, has not only resulted in clean rivers but has also contributed to the development and value of waterfront property. Policies that reduce greenhouse emissions may make certain cities more livable in the summer and might therefore encourage factories to invest in overall plant modernization. These investments, in turn, could increase competition, an economic plus. Thus, if the cost of mitigating climate change turned out to be a good investment, elected leaders supporting such policies would benefit politically. However, if these policies caused companies to close down plants in the United States or move to developing countries with less stringent greenhouse gas regulations, political supporters could suffer at the polls.

An elected leader's stance on the issue of global warming is symbolic of his or her attitude toward environmentalism. A large environmentalist constituency, and an energetic interest group community, is actively engaged in national and international environmental issues. But global warming did not (and, again, excuse the pun) generate much political heat in the United States until 2001, when President Bush came out against the Kyoto Protocol. The paradox is that Bush's opposition may have taken a relatively low-intensity international issue and given it domestic political currency in the United States. He may have had the same inadvertent impact on the issue's standing in the international community. During the 2004 presidential campaign, Bush's position on climate change was part of a set of issues that environmentalists cited as evidence of his anti-environmental stance.

Senators John McCain and Joe Lieberman seized this issue as a way to delineate their own brand of moderate, mainstream environmentalism, and also, perhaps, to tweak a president whom neither was very fond of. The energy industry and other businesses opposed to emission limits lobbied against the McCain–Lieberman Climate Stewardship Act, and the Senate defeated the bill in late 2003 by a 55 to 43 vote. Opposition from the political Right was also intense. Writing about McCain–Lieberman in the conservative *National Review* on October 29, 2003, Marlo Lewis Jr. observed that,

> [The bill would] impose a cap on CO_2 emissions. Carbon dioxide is the inescapable byproduct of the carbon-based fuels—coal, oil, and natural gas—that supply 86 percent of all the energy Americans use. U.S. energy consumption is expected to increase by 34 percent between 2001 and 2020, and carbon-based fuels are expected to supply about 90 percent of the increase. Enacting any variant of [McCain-Lieberman] . . . would be tantamount to issuing a congressional declaration of war on the fuels that power the U.S. economy. Worse, it would establish the institutional framework for a succession of legislative, regulatory, and litigation assaults on carbon-based energy. (Lewis 2003)

Despite strong language on both sides of the issue, climate policy lacks the kind of grass-roots support generated by environmental issues with sustained and visible local effects. Climate change politics is primarily inside the Beltway engaged in by elites, and a factor in the rarefied world of international diplomacy. Its lack of a geographic focus, unlike the geographical significance of a toxic waste site, reduces its salience on the American political agenda. Although the issue is driven by scientific analysis, it is the political perspective that influences how the science of climate change is interpreted. Thus the intersection of science and politics helps to define the issue.

Politics also defines which emission control and carbon sequestration technologies will be used to reduce U.S. greenhouse gas emissions. The political dimension of the issue is partially derived from one's overall attitude toward the role of government. Left on its own to engage in unrestricted, profit-maximizing behavior, industry has no reason to think about long-term effects and reduce its greenhouse gas emissions. Those who believe that the free market alone can best deliver a high quality of life will also resist regulations imposing greenhouse gas reductions. Free market advocates might favor the use of the tax code to induce good behavior, but overall they are not sure about the seriousness of the problem and worry that global warming is merely an excuse to revitalize command-and-control regulation (Victor 2004, 31).

The politics of climate change has domestic variants, as we have seen in the United States, but it is largely an element of political relations between sovereign nations. International relations is an elite politics that reflects the economic interests of nation-states. With globalization, corporations have gained influence over the behavior of multiple nation-states (Levy and Newell 2004, 4). Although national sovereignty remains a powerful force, international regimes (or sets of governing rules and norms) have grown dramatically in the past half-century. This growth emerged, in part, to facilitate international economics and the flow of trade and capital across national borders. Whether the strength of global corporations will enable them to compete with a strong nation-state such as the United States, Japan, or China is unclear. I think that the people who control physical force (armies and police) will still tend to dominate those who control the cash, but this conflict is largely symbolic. Generally these two sets of powerful entities—multinational corporations and nation-states—are acting in concert. Our governing elites are often well connected individuals who move between the public and private sectors, or minimally have strong business and political alliances in addition to social relations across the public and private spheres. Vice President Richard Cheney is a good example of such an elite player, having served as a Cabinet official and also as CEO of Halliburton, the Texas-based construction and engineering firm that serves as a contractor for oil companies and for the U.S. government. These "elites" operate under the assumption that national and corporate interests are either inherently compatible or can and ought to be brought into alignment.

The issue of global climate change has gained status on the international political agenda as the world's political and economic elite have come to slowly accept that climate change is a real problem that could affect the business environment. Swiss Re, one of the world's largest reinsurance companies, for example, has begun to provide insurance against risks associated with climate change, a move prompted by the firm's analysis of the financial risks posed by climate change. In 2003 Swiss Re estimated such financial risks at more than $40 billion a year and expected it to rise to $150 billion a year by 2010. Their risk analysis also led Swiss Re to become a leader in corporate climate policy. Innovations include an internal climate policy, which states:

> Despite advances in research, climate development is and will remain uncertain. Immediate action must be taken nevertheless, as even natural climatic variability carries risks far greater than generally assumed, and man's influence on the climate system will aggravate these risks even further. (Swiss Re 1998)

Swiss Re also intends to become "greenhouse neutral." In October 2003 the company announced that it would launch a ten-year program

> combining internal emissions reduction measures with an investment in the World Bank Community Development Carbon Fund. The voluntary initiative makes Swiss Re the largest global financial services company to set itself the goal to become greenhouse neutral. All Swiss Re locations will participate in the initiative. The programme will utilise the same methodology as Swiss Re offers to clients through its "Greenhouse Neutral" package in partnership with the Commonwealth Bank of Australia. (Swiss Re 2003)

The position of the European Union on climate change and the (slowly) growing recognition of the problem by the U.S. government illustrates the impact of these financial facts on international and domestic climate politics. Poor people will not be the only ones to suffer; the entire global economy could be destabilized. When comparing the costs of reducing climate change to the potential loss of economic activity that could result from global warming, the trade-off seems straightforward.

Global warming is a worldwide political issue with looming potential effects and has gained a growing consensus among our political and economic elite that rapid climate change needs to be controlled. The hope is that the urgency of the problem will reduce political conflict surrounding the issue and focus attention on technology and management concerns. The political dimension of this environmental issue will undoubtedly continue to be intense but, in the end, will probably result in substantive policy to reduce greenhouse gas emissions.

Global Climate Change as an Issue of Science and Technology

The use of carbon-based fuels for electricity, heat, and transportation has changed most people's way of life. It determines the work we do, what we eat, how much leisure time we have and what we do with that time. Our lifestyle is so closely tied to the use of fossil fuels that it would probably be impossible to phase out their use. For that reason, some proposals for reducing atmospheric levels of greenhouse gases focus on removing the gases from emissions and storing them, the process called sequestration, rather than reducing the use of emission-producing fuels. According to Klaus Lackner:

> Climate change concerns may soon force drastic reductions in CO_2 emissions. In response to this challenge, it may prove necessary to render fossil fuels environmentally acceptable by capturing and sequestering CO_2 until other inexpensive, clean, and plentiful technologies are available. . . . Storage time and capacity constraints render many sequestration methods—such as biomass sequestration and CO_2 utilization—irrelevant or marginal for balancing the carbon budget of the 21st century. Even the ocean's capacity for absorbing carbonic acid is limited relative to fossil carbon resources. Moreover, with natural ocean turnover times of centuries, storage times are comparatively short. Generally, sequestration in environmentally active carbon pools (such as the oceans) seems ill advised because it may trade one environmental problem for another. . . . Underground injection is probably the easiest route to sequestration. It is a proven technology suitable for large-scale sequestration. Injecting CO_2

> into reservoirs in which it displaces and mobilizes oil or gas could create economic gains that partly offset sequestration costs. (2003, 1677–1678)

This particular approach suggests that we continue to use fossil fuels for energy until we can replace them with solar- or hydrogen-based sources of energy. A combination of techniques will undoubtedly be applied for increasing energy availability while reducing the presence of CO_2 in the atmosphere.

The development and use of technologies that contribute to global warming is clearly an important dimension of this issue. Human beings have made use of combustible energy since the discovery of fire. Our primary fuel for heating and cooking in the pre-industrial era was wood, which was ultimately replaced by coal; the coal was then used to run turbines to generate electricity. Oil and gas partially replaced the use of coal, and many thought at one time that these fuels would be replaced by nuclear energy. If nuclear power had not posed a dangerous waste and security issue, oil and gas may well have been replaced and perhaps we would all be living in the completely electric houses predicted in the fifties.

Unless we develop a technical solution to reduce CO_2 emissions, our economic growth will suffer and world poverty will probably increase. If the need to reduce gas emissions becomes urgent, powerful nations will likely impose restrictions on less powerful nations, regardless of treaty obligations. The political stability of developing nations would probably not survive a drastic reduction in living standards. As a result, there would probably be no reduction of emission or reductions would be borne disproportionately by weak and poor nations.

Global climate change may only be a precursor of other worldwide environmental issues that will result from large-scale changes induced by the planet's efforts to absorb more than six billion people and their ever growing technology. In their classic work *The Energy Basis for Man and Nature*, Howard and Elisabeth Odum (1981) project the policy impact of the laws of thermodynamics, focusing specifically on the idea that energy cannot be created or destroyed, only transformed. The earth's ecosystems are a closed and interconnected system. When energy is used to perform work, the form of energy degrades and has an impact elsewhere in the system. With global warming, it is apparent that human technology and the level of energy use are resulting in negative impacts worldwide. We should not assume that this is the only impact of our growing population and use of energy; it is simply the first effect we have been able to detect. What will be next? Can we develop the technology needed to measure and mitigate these global impacts?

Considering the technological complexity of dealing with the planet's warming, one may long for "simple problems" such as preventing tank leaks or reducing pollution from waste-to-energy plants. Global problems are especially difficult since we do not have accurate measures of global environmental conditions and have never engineered technology on a global scale. If we devise a technology that fails to mitigate global warming but instead has an unanticipated, negative impact, the consequences could be catastrophic. Even though we cannot ensure success, however, there is no alternative but to try.

Energy consumption and quality of life are inextricably entwined. Social, cultural, and economic development relies on the availability of energy. The genie is out of the bottle and cannot be put back. If we attempt to remove the material well-being people are accustomed to, the political instability that would result could very well lead to the use of weapons of mass destruction. And the effect of those weapons on the planet would be far more damaging than anything that might be caused by global climate change.

The problem of global warming clearly has a number of dimensions. Politics are involved in the effort to regulate the rate of warming as technology continues to develop. But the primary aspect in treating global warming is essentially scientific. Research is needed at a scale resembling the investment in science we saw in the United States during the Cold War. Global warming was caused by technology, and fixing the problem requires the development of new technologies. We need a source of energy that does not generate CO_2 or other forms of pollution.

Global Climate Change as an Issue of Policy Design

We could probably reduce the amount of carbon dioxide and other chemicals released into the atmosphere if we abolish our modern lifestyle along with its technological advances. We could also use the blunt instrument of command-and-control regulation, requiring immediate, massive reductions. Or we could allow developed nations to maintain current levels of emissions and not permit developing countries to increase their use of fossil fuels. But clearly these are all unrealistic solutions. To alter the trend of global warming with few undesirable impacts, we need to design policies that are cost-effective, gradual, and equitable.

Policy designs typically address certain problems while causing others. The U.S. interstate highway system is a good example. While it allowed for faster and safer travel, it also encouraged and indirectly subsidized suburban sprawl. A key goal in policy design is to predict both the direct and indirect impacts of the proposed policy. Despite economists' "assumptions" of certainty, individual and collective human behavior is difficult to foresee. A decentralized, federal political system has the advantage of allowing us to experiment with small-scale pilot projects to gauge the success of a policy. Although a problem as urgent as global warming does not afford a lot of time for experimentation, still various approaches can be attempted simultaneously in different locations. Such experimental policy designs enable us to uncover the approach that works best.

Building on the success of the U.S. acid rain policy, Senators McCain and Lieberman's 2003 legislation proposed a "cap and trade" policy design. Under cap and trade, an overall emission limit, or "cap," is set, and those who emit less than the cap permits can then sell, or "trade," their "pollution rights" to those who cannot easily reduce emissions. This policy helps to ensure the most reduction at the least expense (Victor 2004, 32–33). For certain types of emissions, this approach could be less costly than the pure command-and-control method. There are three other policy designs we could pursue

1. Tax credits or deductions for businesses that demonstrate reduced emissions.
2. Tax credits or deductions for fuel-efficient or nonfossil fuel vehicles.
3. A crash research-and-development initiative to create technologies to mitigate the effect of greenhouse gases, such as carbon sequestration and the development of fuels or energy technology with little or no emission.

The international nature of the problem requires different design elements for nations at various stages of economic development and an enforcement mechanism for noncompliant states. As noted, when the largest single emitter of greenhouse gases, which also happens to be the most powerful nation in the world, refuses to participate in an international governance scheme, policymaking is undermined. No amount of policy design finesse and creativity can overcome the political setback caused by the U.S. rejection of the Kyoto Protocol.

Incorporated in the Kyoto Protocol are certain design elements to ensure equity and impact. Policies proposed by the European Union and McCain-Lieberman are examples of reasonably sophisticated, "first-generation" policy designs that, if adopted, could begin the learning process needed to achieve emission reductions. At this writing, the United States lacks the national political will to begin this policy process, even in the short run. Nor do we know if the reductions that can be achieved without economic disruption are sufficient to resolve the problem of global climate change. A technological fix that would allow continued economic growth without environmental damage is key to addressing this issue.

Global Climate Change as a Management Issue

If the need for new technology is at the heart of the climate change issue, the management dimension is to develop the capacity to invent and then deploy such technology. The organizational capacity required to develop a new source of fuel or a practical form of carbon sequestration would be on the scale of NASA's moon project in the 1960s. Shifting from fossil fuels to other forms of energy would also require considerable organizational and social learning. For these reasons, the climate issue will be dominated by management considerations at some time in the future. For now, however, it is too early, in terms of policy, program, and technological development to consider management a predominant dimension of the problem.

Organizational capacity is currently needed to continue research and scientific development on the impacts of climate change, to work with nation states in developing strategies to reduce greenhouse emissions, to monitor corporate behavior, and to encourage emission reductions. At this stage of policy development, organizational learning is a high priority. We need to improve our understanding of the dimensions of the problem and to stimulate the behaviors needed for reduced emissions. We have many models to draw on to build institutions that promote organizational learning and the development of science and technology to reduce global warming. That is the easy part. But once we learn what is required, we then need to build nation-specific and global institutions that can implement those solutions. The difficulty of that task is unprecedented.

Summary of the Multiple Dimensions of Global Climate Change

As environmental policy evolves, the ecological problems we face become increasingly complex. Climate change, the first truly global environmental issue, challenges our political institutions. These institutions were designed to deal mostly with local problems and lack strong cross-national and international governance mechanisms. A principle objective of government is to maintain public security, sustenance, and safety. Historically, when threats to security became broader geographically, governments in turn grew to represent more extensive areas. The emergence of the United States of America and the European Union are examples of this phenomenon. If threats to our security are global, history tells us that government institutions on a worldwide scale may soon follow. Although a world government is difficult to envision, the emergence of a global economy and global environmental problems may be the start of demands that could lead to a global political structure.

Our capacity for self-destruction may also necessitate new types of global political institutions. We obviously cannot predict how global political institutions, processes, and practices will evolve, but they may indeed be essential to manage unprecedented problems such as global warming. New political institutions will not be needed, however, if nation-states and global corporations can meet the unparalleled threats now facing us. Institutions currently in power clearly do not want their authority threatened by a new institutional arrangement, and therefore have every incentive to respond to these new demands. Enlightened self-interest, however, is considered enlightened precisely because most powerful leaders define their self-interest in narrow and conservative terms and do not see the need for change until it is too late. Franklin

D. Roosevelt's welfare state, a response to the Great Depression, may well have preserved the power structure in the United States. FDR, however, was not a typical governing elite. If those in power today want to remain in charge, they will need to implement the principles of sustainable development, that is, protect the environment and reduce global poverty. Can they rise to the challenge?

The world's political and economic elite have gradually accepted the facts of global warming. But can they transform that acceptance into resources to develop the technologies that will enable us to maintain our current lifestyle without generating more greenhouse gases? A technological optimist would answer yes. A skeptic would perhaps predict catastrophe. In a sense, we are back to Malthus, needing to project the future based on trends. Malthus predicted that we would overpopulate the planet and then be unable to produce enough food to feed everyone. But he did not account for technological innovation in the production and distribution of food. Will today's science and technology come to our rescue with innovative ways to resolve the climate issue? The jury is out.

Conclusions

Climate change is a reflection of the power of our technology and the difficulty of controlling its impact. As a society, we depend completely on technology and are entranced by its magic. But how much more technological do we want our world to be? As a child, I watched "The Jetsons," an animated TV show that featured a family of the future living in a totally technological world. The dog took a walk on a treadmill, cars flew through the air, and the maid was a robot. The natural world did not exist. People lived in the sky and got their food from dispensers on the wall. When I first began to study environmental policy and values in the mid-1970s, I recalled that show and wondered what had happened to the Earth below. The Jetsons never saw a tree, a river, a mountain, or an ocean. George Jetson worked three hours a day, and material comfort was assured. Are we capable of creating such a world, and would we want to? The Wealthy Texas oil magnate and philanthropist Ed Bass funded the Biosphere II experiment in Arizona to see if a self-sustaining ecosystem was possible. The results showed that, more than forty years after the Jetsons' technological world was conceived, we still lacked the science and technology needed to operate our own biosphere. The Biosphere II could not provide enough food and oxygen to support human life. We still needed the original biosphere, the planet Earth.

Our way of life in America is not that of the Jetsons, but we do enjoy a lifestyle that is an energy-dependent, technological marvel. Visit a southwestern suburb in the United States. Drive your three-thousand-pound, air-conditioned SUV on an eight-lane interstate highway to your four-thousand-square-foot, climate-controlled home, your kitchen laden with modern appliances that range from a subzero freezer to an automatic icemaker. Enter your den where you find your computer and World Wide Web–based entertainment system that can record entertainment for you to view at your convenience. Swim in your climate-controlled pool, or water your garden, with water that has been pumped from a mountain range more than a thousand miles away. Clearly, although we do not yet have the technology of the Jetsons, we still have dreamlike technology and, with it, parks, forests, oceans, and other natural environments to play in, relax in, and marvel at. But technology comes at a price and, with global warming, the bill for our technology is coming due.

Here we leave behind the discussion of sample environmental problems and return to a consideration of the preliminary framework for understanding environmental policy issues presented at the start of this book. Has the framework given us a more thorough, multidimensional understanding of these environmental issues? We explore this question in the final section of the book, "Critiquing the Framework."

Part III

Critiquing the Framework

Chapter 7

What Have We Learned from the Framework About Environmental Problems, and What Else Do We Need to Know?

Summary of the Framework

Parts 1 and 2 examined environmental issues through a framework that provided a common set of questions:

- What is the value dimension of the problem?
- What aspect of our lifestyle led to the problem? Can the problem be solved by changing the way we live? And is that possible, or are these behaviors too central to our culture and value system to be changed?
- What political issue does the environmental problem pose, and how did it get on the political agenda? What political, economic, and social forces created the problem? What political and institutional arrangements might help to solve the problem? How did the issue's definition on the political agenda influence the shape of the program designed to solve the problem? Is there political consensus behind the definition of the problem and the proposed solutions?
- Was the problem caused by the development and use of new technology? Can new technologies be developed to control the negative impacts?
- Can effective methods be developed to inspire corporate and private behavior to reduce or mitigate environmental damage? Have such proposals been developed, and have they achieved status on

the political agenda? What needs to happen in order to promote effective policy designs?
- Do we currently have the organizational capacity to solve the problem, or does it need to be developed?

In addressing these questions I highlight elements of the issues and provide a more nuanced description of each problem. My effort here is not to develop an objective description of the environmental problem but rather to understand it as a policy issue. To do so, several questions need to be addressed:

- How has the problem been defined by the political process?
- Which aspect of the problem dominates?
- Why is the problem seen in one way and not another?
- What is the relationship of proposed solutions to the nature of the problem?
- Which part of the problem is being targeted for response or solution, and why?
- What is the relationship of the proposed solution to existing or potential organizational capacity?
- Do we know how to do this work? Does anyone else?
- Does the work require new technology, or can we use the materials and techniques that are sitting on the shelf?

The framework explicitly deconstructs the environmental issues into component parts and examines each aspect separately. Analyzing each element related to a given problem helps to illuminate its most important features. One aspect of the framework, discussed later on, is the need to examine how the various dimensions of the problem interact and to analyze the impact of one element upon another.

Applying the Framework: What Did It Tell Us?

By using the framework to identify the main element of each environmental problem, we can better allocate the necessary resources and energy into appropriate channels to solve the problem. New York City's solid waste issue, for example, does not require new technology or management capacity; it is largely a political issue of facility siting.

Underground tank leaks, on the other hand, are essentially a management issue, assuming that we continue to use internal combustion engines in our vehicles. The example of toxic waste cleanup featured important dimensions of each element of our framework. It is a complex issue with a fairly balanced multidimensional definition.

In fact, each element of the framework sheds some light on a problem's definition and proposed solution. In some cases one element of the framework dominates, whereas in others no single element takes precedence. For a problem to become a public policy issue, by definition it must have political dimensions. All the environmental problems we have examined here seem to share a common origin: the value system that has led to our resource-consumptive lifestyle.

Sometimes one's expertise in a particular discipline leads one to focus only on a certain aspect of an issue and assume that the entire issue is being analyzed. Often people who do not understand the policy process make foolish assumptions about what that process can accomplish. Environmental scientists examine the science of an environmental problem, identify its cause and effect and possible solutions, and then lament the messiness and irrationality of the policy process for acting incrementally and in half-steps. Economists apply their often elegant, but all too imperfect, models of individual and collective human behavior to these same problems and identify tidy policy designs to solve the problem—if only those foolish public servants who make policy could understand and carry out these designs.

Formulating and implementing policies are complicated, messy, and unpredictable activities. The framework proposed here offers details on how the policy process differs based on various factors. When assessing policy formulation and implementation, no human activity can be viewed in isolation. Context matters. Geography matters. Culture matters. A policy can be designed for fees to be charged for residential garbage pickup in Phoenix, Arizona, where most people live in private homes or garden apartments and one can tell whose garbage is whose. In New York City, on the other hand, most people live in apartments and garbage is placed in a common area in the basement, making it difficult to assign fees to households. Thus a policy design that works in Phoenix would not work in New York City.

Even if waste-to-energy plants were determined to be the most environmentally benign and cost-effective means of managing garbage in New York City, such plants might not be the most viable solution since siting them in New York City is highly contentious. Community opposition and intense political heat, typical factors in the NIMBY or "not in my backyard" syndrome, often cause elected leaders to shy away from proposing these types of facilities.

In other words, politics, management, values, and technology matter. Policy is complicated, because human beings and their needs and interests are complicated. The framework proposed in this book allows us to examine some of the complexities of these four issues. The sections in chapters 2, 3, 4, and 5 summarizing the multiple dimensions of the framework is a recognition of the need to deal with interactive effects within the framework. For example, the management dimension is influenced by the nature of the technology being managed. If underground tanks contained a substance less toxic than gasoline, the organizational capacity needed to handle the fuel could be less sophisticated. On the other hand, if we were trying to handle reactor fuel for a nuclear power plant, it would be more toxic and far more difficult to manage than gasoline.

The framework presented here encourages the analyst to consider, and roughly prioritize, a broad range of critical issues. Indeed, a highest-priority dimension for each of the environmental issues discussed can readily be expressed:

- The dominant aspect of New York City's waste issue is *politics,* especially the politics of siting facilities.
- For underground tanks, assuming the technology of autos continues to require gasoline, the predominant element is *management* and organizational capacity.
- Toxic waste cleanup has a variety of important dimensions, but since we still do not know how to detoxify land, the need for new *science and technology* is the dominant dimension.
- Similarly, even if we could muster the political will and institutional arrangements to reduce or sequester emissions of greenhouse gases, we still do not have the technology to do so.

This suggests, perhaps, that there is an identifiable sequence to formulating and addressing environmental issues. The first step is to define the problem and then determine the science needed to measure its dimensions. Typically the problem is a function of our values, whereas its definition relies on the politics surrounding the issue. For example, with regard to leaking underground tanks, our attraction to freedom and mobility is what created the problem. That we focus on toxic waste cleanup rather than controlling the very creation of toxic waste is a social construction. We define the policy problem in terms of what we consider to be the most significant dimension of the problem. In other words, our understanding of the issue depends on the scientific paradigm or worldview that shapes our choices of problems and potential solutions.

Whether the issue will reach the institutional agenda, and how it will do so, is shaped by political forces. We need to develop technology to reduce, eliminate, or clean up the damage. Developing this technology may require political will to address the problem. It is possible, however, that the process of technology development could be organized wholly within the private sector, thus reducing the need for political support. If the public sector develops the technology to address the problem, we will then need political approval to obtain the resources to implement the technology. If and when political approval is secured, we then need to develop the organizational capacity to manage the technology.

Applying the framework to these environmental issues helps us to define and understand these problems more precisely. We then can develop a realistic understanding of the remedial policies we need and how quickly we can develop and implement them.

Values and Environmental Policy

Although all dimensions of the framework are relevant to most environmental policy issues, the role of values is most fundamental to certain issues. All four issues examined in this book are the result of our lifestyle and the consumption required to maintain it. Applying the framework reinforces the centrality of the value issue. This brings me back to my training in environmental policy and the work of Lester Milbrath, my dissertation chair, on environmental perceptions and values. Milbrath's concept was straightforward. The cause of environmental problems is our system of values. The "dominant social paradigm" (DSP) in our society relies on technology to solve environmental problems and is based on the belief that humans should dominate the earth, and use and consume its resources for their own gratification.

Writing in the early 1980s, Milbrath described a New Environmental Paradigm (NEP) that "does not . . . renounce all technology, all industrial production, all growth, or all material goods . . . [instead] advocating thoughtful consideration of where we are going, careful and subdued production and consumption, conservation of resources, protection of the environment, and the basic values of compassion, justice and quality of life" (Milbrath 1984, 14).

In Milbrath's view, in order to change our value system and transform our society's dominant social paradigm to a new environmental paradigm, we must first understand this value system in great detail. To accomplish this, we need to study the ways that people perceive the environment. Then we need to educate people so they understand that the earth's resources and ecosystems are finite and fragile, that they are doomed under the dominant social paradigm and require the adoption of a new environmental paradigm.

The problem with Milbrath's formulation is not his analysis of the problem but rather the solution he proposes. It is not realistic that those with the power to continue their level of consumption will change their behavior simply because it is not sustainable. But although the solution Milbrath suggests may be impractical, his analysis of the root cause of the problem is persuasive. That a deeper understanding of people's environmental perceptions could lead to a better understanding of environmental values is true. The perception that leads people to drive SUVs and the principles that allow toxic waste to be stored on private property contribute to the destruction of the environment.

According to Milbrath, without a large-scale change in our worldview and the adoption of a new "soft-path" paradigm of sustainable resource use, the planet will cease to produce the materials needed to sustain human life. Small-scale, remedial, and incremental changes in environmental policy will be too little, too late.

Today, twenty years later, elements of environmental awareness are certainly more widespread, as is support for environmental policy goals. But the Western style of material consumption has not been replaced by a more environmentally oriented paradigm. People worldwide strive to achieve the consumer-oriented lifestyle that persists in the West and in Japan. The economic interests that shape these values and transform them into desires and needs have only grown stronger in the past two decades. Moreover, satellite TV, thousands of cable channels, and the worldwide web have provided corporations with many more avenues to disseminate their messages than were available in the 1980s.

Environmental quality itself has become a consumer good. Ecotourism, second homes, camping, hiking, and other travel-related activities have brought more and more people into increasingly fragile environments. We value the environment more than ever and want to ensure that we can benefit from clean air, unpolluted water, mountains, and other natural settings. The NIMBY syndrome is partly a reflection of the popular desire to keep environmental insults away.

The framework presented in chapter 2 always brings us back to the fundamental, yet challenging issue that every environmental problem we face is grounded in our value system and our view of how the world works. The capacity of economic interests to drive and manipulate the shape of environmental values over the past twenty years has been impressive and irrefutable. Our need to circumvent this basic fact is part of my motivation to develop a multifaceted framework for understanding environmental policy. If we cannot change the fundamental values that create environmental problems, perhaps other levers can be used to provide policies that will remedy these problems without requiring us to change our deep-seated beliefs and lifestyle.

To develop effective environmental policy that reduces pollution and leads to sustainable development, we need to examine and understand the current political reality and the influence wielded by those with vested economic interests. People with those interests oppose fundamental change that threatens their power. Therefore alternatives must be found to bring about the massive paradigm shift discussed by Milbrath. A transformation of values is a very slow process, and social scientists do not yet have sufficient knowledge or understanding as to what stimulates a shift in values and what the impacts will be. Incremental change, as Milbrath argued, may be too slow to solve the rapidly developing environmental problems we face. But the mistakes that may accompany a slow process of change will also occur incrementally and perhaps be more readily resolved. If the price of slower progress will avoid catastrophic mistakes, that is surely a price worth paying.

Without question, technology develops more quickly than public policy. So it is possible that environmental policy will develop too slowly to solve our problems, and we may witness the slow or even rapid deterioration of the planet that sustains us. These are the realistic constraints under which we operate. Again, while I acknowledge the importance of understanding the value basis at the root of environmental problems, as Milbrath stressed, I do not believe that a fundamental change in our value system is a realistic solution to environmental problems. Instead, we need to develop policy approaches that acknowledge the deep roots of our behavior and the critical need to solve our problems but without a fundamental change in the way we live.

Politics and Environmental Policy

We know that all environmental problems do not share equal levels of political saliency, do not occupy the same spot on the political agenda, and do not generate the same degree of political conflict. There is a great deal of symbolism in environmental politics. As the United States has de-industrialized, environmentalism has shed some of its anti-blue-collar image. There is little political cost in being labeled an environmentalist. In fact, such labeling can be seen as a political boon. Still, some elected leaders continue to cling to the trade-off between economic development and environmental quality. At the same time, an increasing number of leaders understand that the relationship between economic growth and environmental protection is far more complex than a simple one-to-one choice. Moreover, support for environmental protection cuts across all shades of the political spectrum. The largest environmental group in the United States, for example, the National Wildlife Federation, includes a large number of conservative hunters. Polls indicate strong support for environmental protection among both liberals and conservatives.

All four cases examined here show that an increasing number of environmental issues are reaching the political agenda. In the past half-century, institutional structures have been created to address environmental issues at all levels and in all branches of the U.S. government. Some hot-button issues that dominated the media have become routine. Other concerns, such as underground tanks and sewage treatment, which started as local issues, became visible and made it to the national political agenda for a time, only to become local again and subject to routine county- and municipal-level action, even though they now had federal regulatory standards and grant funds attached.

Under the current Bush administration, there has been a renewed politicization of environmental issues as a result of the administration's ideological fervor. At the same time the political center has become intensely pro-environment. Thus some of the controversy associated with environmental politics will probably evaporate in the long run. The center-oriented dynamic of the U.S. political system is driving anti-environmentalism to political extremes. Climate change and toxic waste are difficult issues to solve, but very few people think they can be ignored. Our increased experience with environmental issues in the policy process, and the likely emergence of new environmental problems, can be expected to result in a continual flow of new and redefined environmental issues onto the political agenda. The contentiousness of this policy process will vary, depending on the complexity and danger posed by the environmental issues as well as the resources needed to address them.

The increased resources being allocated to environmental protection has made both the public and the ruling elite view the politics of environment no longer as a critical, extraordinary matter but instead as a routine political dialogue governing the decision processes of a routine public administrative function. Waste management, toxic waste cleanup, water filtration and supply, sewage treatment, and air pollution control are now seen as routine costs of doing business in the U.S. Just as the government builds roads, schools, and hospitals to maintain a standard level of public well-being, most people simply assume that it also keeps the environment safe. These cases show a pattern of political issues taking the normal path through the issue attention cycle from crisis and extraordinary politics to the routine of consensus-oriented day-to-day political dialogue. Even the latest and most critical issue of global climate change shows signs of reduced conflict and growing consensus.

Science, Technology, and Environmental Policy

The issues discussed in this book are all exacerbated as well as potentially solved by technology. Many of the technologies needed to resolve these problems are already well developed and accessible. Waste management and underground storage tanks pose few technical challenges. The past quarter-century has seen the achievement of promising technological advancements in toxic waste cleanup. Global climate change, as the most recent issue, has obviously had the least time to develop technical fixes.

The contention surrounding environmental politics is largely a function of the cost and complexity of the technology available to solve these problems. A high degree of scientific literacy is needed to fully comprehend environmental policy. A knowledge of chemistry and biology, as well as ecology, is essential to understanding the relationship between pollution and its effects on the ecosystem and on human health. Engineering principles are also involved in developing cleanup strategies.

In my personal experience with environmental decision makers it appears that they, like me, have little formal scientific training. This lack of knowledge and understanding substantially reduces our ability, as policy makers, to address environmental problems effectively. Environmental leaders are predominantly trained in business, law, and public policy, and will need to acquire scientific and technological training if they are to be truly effective in resolving today's compelling environmental issues.

The Design of, and Economic Impact on, Environmental Policy

Environmental policy has benefited from the fact that it is a relatively new area of regulation. Whereas the initial laws regulating air and water quality and hazardous waste were purely of a command-and-control nature, most subsequent regulatory laws have benefited from creative policy designs. Most of these designs have been based on modern economic concepts and have consequently provided more cost-effective means of changing environmentally damaging behaviors.

We have come to understand and accept the realistic constraints that limit elegant environmental policy designs. "Cap and trade" policies, for example, have been used to reduce the sulfur dioxide pollution that causes acid rain. Because of the way sulfur dioxide is transported through air, higher concentrations of sulfur dioxide in one geographic area can be offset by lower concentrations in another location. Similar proposals for mercury are not feasible because of the way that mercury is transported and also because of its tendency to become concentrated in "hot spots" (Barnhardt et al. 2004). The phrase "all things being equal" is often used by economists as a way to dismiss variation that may interfere with the functioning of a model they are proposing.

When attempting to manage ecological systems and chemical contamination, however, all things are rarely equal and assuming so can lead to dangerous conclusions.

As Walter Rosenbaum has observed, "The impact of environmental policy on the economy is a constant preoccupation of environmental regulators and the regulated." (Rosenbaum 2005, 101–102). The focus of environmental policy design is always to minimize economic costs while maximizing economic and environmental benefits. In the United States, the move from a manufacturing-based economy to an economy based on services and information has changed the input side of the design equation and reduced the economic costs of some environmental policies (ibid.).

Still, the search for imaginative policy designs should not be halted because of the difficulty of the task. Just as engineers and scientists can invent new technologies, so, too, can social scientists develop innovative ways to organize and influence individual, corporate, and social behaviors. We need to take a modest approach to these design tasks, however, to avoid creating the same types of unanticipated impacts facing our colleagues in the sciences.

Implementing Environmental Policy

The implementation of public policy requires specific behaviors by specific people. Typically these behaviors are coordinated actions taken by groups of people in different organizations. No matter how elegant the policy design, no matter how many incentives are in place to encourage these behaviors, if people do not perform the tasks, nothing happens. As one of my students once told me, "People are always, well, so human!" And he elaborated: "People are unpredictable, difficult to understand, and sometimes unmanageable." Effective management may seem simple and is often assumed to already be in place, but it is frequently the missing piece of the policy puzzle. The absolute failure to implement emergency response in the aftermath of Hurricane Katrina in September 2005 demonstrates graphically what happens when organizational capacity is taken for granted. In this case, local and state governments were quickly overwhelmed by the unprecedented catastrophe. While the federal government possessed the capacity to provide emergency response, ineffectual leadership failed to deploy that capacity. No decision was made to let people suffer and die, but inattention to communication and management had that effect. The stated policy of the U.S. government during an emergency is to provide assistance to its citizens. This policy was not implemented after Hurricane Katrina.

I do not mean to imply that typical organizational behavior is no behavior at all. Indeed, for every person who takes a two-hour coffee break, dozens of other individuals and groups are performing at extraordinary levels. We saw such activity in New York City after the destruction of the World Trade Center. The site was cleared months ahead of schedule, and the subway tunnels below were repaired in record time. Workers at the site saw their efforts as a testament to the people who died at the hands of terrorists and so, driven by emotion and a sense of duty, they worked harder, better, and longer.

Management is a critical dimension of environmental protection and a necessary factor in understanding environmental policy. The methods used to clean up toxic waste sites can be seen as the transfer of standard operating procedures used in the construction of sewage treatment plants. Those treatment plants, in turn, owe their lineage to the construction of dams by the Army Corps of Engineers. The procedures used to clean up oil spills in the Gulf of Mexico were the basis for the routines followed by Superfund's emergency response program. The persistence of leaking underground storage tanks can be explained as a failure of organizational capacity.

Understanding environmental policy requires that we look at the specific elements of the policy that actually become real. We need to understand the authority, funding, personnel, skills, and experience that organizations require to perform their work. We must examine the organizational routines, the actual labor processes, the typical mistakes, the work process improvements, and the coordination of efforts that form the behaviors we call "work." We need to understand precisely what outputs organizations are producing and the impacts of those products. Even though we issue effluent discharge permits, we need to know how much pollution is actually coming out of those pipes. Is the idea of issuing pollution permits even working? Do we need to come up with something else? One cannot manage an activity unless it can be measured. Without measuring the outcomes of a policy, we cannot determine whether the changes made in organizational behavior are making things better or worse (Cohen and Eimicke 2002, 179–183).

The cases reviewed here reinforce the critical nature of management in warding off environmental hazards. When management is not yet an issue, as in our current problem of global climate change, we have no real, functioning public policy. Once a climate change policy is in place, the global nature of that policy will probably require that we learn from the standard operating procedures of multinational corporations as well as those used by multinational nongovernmental organizations. Coordinating and verifying behaviors from multiple nations will likely be complicated in ways we can barely imagine.

Limitations to the Framework and Possible Modifications

The framework this book proposes is clearly based on the assumption that environmental problems are related and share common properties. Institutions concerned with the environment, such as the U.S. EPA and its counterpart agencies at state and local levels are also founded on the notion that environmental issues are connected. While the framework poses a set of questions that adds to our understanding of the issue, it is less systematic about the interactive effects. For example, we could combine science and technology with management issues since the degree of technological complexity places demands on organizational capacity. Similarly, value issues are often reflected in political dynamics, and so these dimensions could be combined. Following that logic, we could chart every possible combination of causal pathways by explicitly examining every feasible interaction between the elements of the framework. Perhaps a number of cells in that matrix may be useless or even silly, but such an analysis would be more comprehensive than the explorations presented here.

Other modifications to the framework would be to add elements or else disaggregate the existing elements into narrower subfields. Any efforts to illuminate the various aspects of the policy problem, and avoid confusion, is welcome. The areas I have included in the framework relate to my understanding of the sources of environmental problems and the factors contributing to the way we define and work toward resolving these problems. What is essential is that the inquiry is broad enough to address the following underlying questions:

- How do we define the problem?
- What human activities have created or exacerbated the problem?
- How is the problem understood by the political and economic elite and by the public at large? Do the decision makers understand the problem and its impact?
- Do we know how to solve the problem? And, assuming we do, what are the constraints on implementing the solution?

The reason for asking these questions is not dispassionate, objective inquiry. Our goal in using this framework is to understand as much as we can about the problem being addressed and the approach being taken so that we can improve environmental policy and ultimately sustain the planet. The final chapter discusses possible ways to encourage the development of more effective environmental policy.

Chapter 8

Conclusions

Improving Environmental Policy

Our goal is to improve environmental policy, which does not imply that we are doing a bad job. Quite the contrary, our environmental agencies have made impressive progress in addressing environmental issues and are often unjustly criticized. Our society, academic community, and government and private institutions have learned an enormous amount about our environmental problems in the past thirty years. The cases described in this book illustrate the evolution of various environmental issues and demonstrate that social learning and institution building are central to developing effective environmental policy, even if it is often a case of two steps forward, one step back. Despite this progress, however, two broader questions need to be addressed: Are we adding people and pollution faster than policy can create and implement ways to control their impact on the planet? What can be done to develop a policy in response to our growing set of environmental issues?

Accomplishments of Environmental Governance

Thirty years of incremental progress adds up. As noted in chapter 2, GDP growth in the United States has been decoupled from pollution growth. Our rivers, streams, bays, and oceans in some places are cleaner than they were in 1970. The growth of waterfront development is an indicator of that progress and has created a strong pro-environment interest group of waterfront residents. At the same time, waterfront development has created a population vulnerable to natural disasters, as their homes are located in fragile ecosystems; this population also poses a threat to the local environment. Still, air pollution is less toxic in some areas than it used to be, and we have managed to detoxify some hazardous waste sites.

Most important than these specific program accomplishments has been the general increase in environmental awareness and knowledge in the United States and throughout the developed world. This increased awareness has not always resulted in effective environmental policy, but it is a necessary, though insufficient, condition for the development of such policy. The public knows that the environment is threatened. They see it in their daily lives, and support policies and programs to improve environmental quality.

To some observers, the reason we do not have effective environmental policy is that entrenched interests, protecting their wealth and power, prevent environmental issues from reaching the political agenda and block certain policy approaches to environmental protection (Rosenbaum 2005, 34–38). We damage the environment by creating massive consumer demand in the public through messages sent via electronic media. Some believe that these demands are stimulated by economic elites who profit from this consumption (Holt and Schorr 2000, xi).

I am not certain that this is the case. I do not think that the mass consumer society is wholly created by advertising. I believe that people like this more consumptive lifestyle, and that one's choice of lifestyle is not so easily subject to media manipulation. Air conditioning on a hot day feels good. A ride in a luxury car is more comfortable than a ride in a compact car, and certainly it is easier than walking. We do not need to be told that we like large, comfortable personal spaces; we can figure that out ourselves. The shape of our consumption is manipulated, but material goods are seductive. In other words, we do a good job of convincing ourselves to consume material goods; we do not need advertising to inspire us (Twitchell 2000, 288).

What is changing in the West is that this desire to consume is increasingly coupled with a basic understanding of ecological fragility and limits. Consumption continues to increase, but so, too, does environmental awareness. This increased awareness is the result of decades of work by scientists, policy makers, educators and the media. This is a very important accomplishment, and all other achievements in this field are derived from this one. People think about the environment, even if they do not always act as if they care about it.

A second major accomplishment has been the development of organizational capacity in the public and private sectors to protect the environment. When the EPA was formed, it had a staff of about 3,000 employees. Today the EPA employs more than 18,000 people, and its budget has grown from about $2.2 billion in 1970 to a little over $7.5 billion in 2004.[1] Much of the additional funding has paid for contract staff, and thus the actual growth in staffing is much greater than the 600 percent increase indicated by EPA personnel data.

A third major accomplishment has been to institutionalize thinking and sometimes action about environmental protection in government and business. The requirement of the 1969 National Environmental Policy Act (NEPA) to include environmental impact statements in large-scale development projects is often viewed as ritualistic and frustrating, but it *has* forced people to think about the environmental impacts of development. Lynton Caldwell, founder of the field of environmental politics and the principal author of the environmental impact statement requirement in NEPA, stated in the late 1990s that he considered NEPA largely unimplemented (1998b). In his book, *The National Environmental Policy Act: An Agenda for the Future,* Caldwell observed that the full institutionalization of NEPA needed to be seen only as a hope for the future. Caldwell testified before Congress about NEPA's impact after thirty years and stated:

> Few statutes of the United States are intrinsically more important and less understood than is the National Environmental Policy Act of 1969. This comprehensive legislation, the first of its kind to be adopted by any national government, and now widely emulated throughout the world, has achieved notable results, yet its basic intent has yet to be fully achieved. Its purpose and declared principles have not yet been thoroughly internalized in the assumptions and practices of American government.... Through the judicially enforceable process of impact analysis, NEPA has significantly modified the environmental behavior of Federal agencies, and indirectly of State and local governments and private undertakings. Relative to many other statutory policies NEPA must be accounted an important success. But implementation of the substantive principles of national policy declared in NEPA requires a degree of political will not yet

> evident in the Congress or the White House.... Three decades since 1969 is a very short time for a new aspect of public policy—the environment—to attain the importance and priority accorded such century-old concerns as taxation, defense, education, civil liberties, and the economy. The goals declared in NEPA are as valid today as they were in 1969.... The most important and least appreciated provision of NEPA is the congressional declaration of national policy under Title I, Section 101: "that it is the continuing policy of the Federal government... to create and maintain conditions under which man and nature can exist in productive harmony, and fulfill the social, economic, and other requirements of present and future generations of Americans. (Caldwell 1998a)

To Caldwell, the environmental policy "glass" may be half-empty. But although he laments the lack of attention paid to environmental policy, he acknowledges that environmental impact analysis has affected the behavior of governments and changed the way they assess the costs and benefits of development projects.

Sitting here in New York City, where the powerful public official Robert Moses once built highways that destroyed neighborhoods and ecosystems with impunity, I think Caldwell does not give himself enough credit for an institutional innovation that has changed how governments and industry do business. While sometimes it is lip service, and some of the people building roads, bridges, factories, and housing do not really care about environmental quality, nothing large is built in the United States today without an assessment of its effect on the environment. This is a necessary although insufficient condition of sustainable development. Without an impact analysis, we cannot even have a conversation about the costs and benefits of development.

A more detailed analysis of the state of environmental policy in the United States was presented by Daniel Mazmanian and Michael Kraft (1999) in their analysis of the evolution of environmental policy since the 1970s. In their assessment of thirty years of environmental policy development, they identified three epochs of the environmental movement (10, table 1.1):

1. *Regulating for environmental protection (1970–1990)*: D.C.-based command and control regulation of end-of-the-pipeline pollution with a focus on air and water pollution.

2. *Efficiency-based regulatory reform and flexibility (1980–1990s)*: State- and local-oriented use of markets and cost-effective analysis to encourage companies to internalize pollution control with an added emphasis on toxic chemicals.
3. *Toward Sustainable Communities (1990 on)*: "Bringing into harmony human and natural systems" (10). The objective is a locally oriented effort to ensure that our actions are environmentally sustainable.

Mazmanian and Kraft further observed that:

> The concern of the third epoch goes well beyond prescribing regulations for cleaning up pollution or conventional cost-benefit analysis of their effects. What is being asked for is a method of gauging the multiple ramifications of an action—rule, regulation activity—within a large and complex array of possible effects, in the near term and far into the future. The level of scientific and technical data, understanding of ecological processes, and analytical capability needed for this kind of assessment is greater than ever before. (28)

This description of three epochs can be contrasted to the three approaches of my own view of the evolution of environmental policy since 1970: the environment as (1) an aesthetic issue, (2) a human health issue, and (3) an issue of the health of the biosphere. Some of those who have participated in or thought about environmental governance since the 1970s tend to see an evolution of policy objectives, processes, organization, and outcomes.

The observed evolution has been toward increased effectiveness, greater efficiency, and a broader scope of environmental policies. In some respects, that expansion is one of the major accomplishments of these first years of environmental governance. As Caldwell observed, we seem to be in the process of adding the function of environmental protection to the set of traditional government services. We are in the process of learning what policies and programs seem to work and at the same time evolving new institutional forms as the scope of the problem changes and expands.

Mazmanian and Kraft's focus on community-based environmental governance is echoed by Edward P. Weber (2003). Weber examines efforts in the rural western United States to develop local means of resolving conflicts over the use and protection of natural resources. This "grassroots ecosystem management" (GREM) approach is an attempt to bring environmentalists, developers, loggers, business leaders, and government officials together to develop compromises that allow for both environmental protection and business use of natural resources. Weber focuses considerable attention on the need to reconcile this new form of governance with the requirements of democratic accountability. The "place-based" advisory groups formed under GREM have no formal authority, and yet they have enormous legitimacy and influence at the local level. He effectively demonstrates that these negotiations at the local level do not pose a threat to representative democracy and may in fact be a healthy corrective to the tendency for general principles established at the national level to be distorted and misapplied at the local level. Weber's work details the need for place-based mechanisms for interaction and negotiation to augment and complement national rule making.

This local form of environmental governance is not simply seen in rural communities. Community-based environmental nonprofit organizations are growing in cities throughout the United States, as have more participatory local land use planning processes conducted by zoning boards, planning agencies, and town councils. As local elected leaders have faced the NIMBY syndrome and term limits, they have increasingly sought to consult with local communities on economic and real estate development projects. They do this for political survival, not for altruistic reasons. Efforts to mitigate community and environmental impacts have become a routine part of business operations. Typically developers set aside between 5 and 10 percent of the project cost for amenities for local communities. A striking example of this tendency is the Riverbank State Park that was built on top of the North River sewage treatment plant in the West Harlem–Washington Heights area of New York City. This park features an indoor swimming pool, ice skating rink, community meeting rooms, and a soccer field. If the state had not agreed to these "side payments" to the community, the political opposition to siting the plant might have prevented its development.

These new community-based institutions are only in their infancy, but they show signs of development and maturation. Use of the Internet has reduced the cost of communication between neighbors in urban areas and between developers, elected leaders, and communities. This enables groups to meet both "virtually" and in person, thus providing real-time input during the development process. I do not argue that these groups work perfectly or that the forces of economic development do not try to use their power to force change on reluctant communities, but the days are long gone when a Robert Moses, or his counterparts in other cities, could simply brush aside community concerns and build at will.

I mention this because at one time development was seen as inherently positive. *The Urban Villagers* by Herb Gans (1962) and *Death and Life of Great American Cities* by Jane Jacobs (1962) are seminal works identifying the negative impact of development on communities and their culture. In the early twenty-first century the burden of proof for development has been reversed in many locations. In the 1960s the opponents of development had to prove that a project might cause harm. Today developers must often convince a skeptical community that the benefits of a project outweigh its costs. This is a sea change, and we should not underestimate its political impact.

More than thirty years of environmental institution building, social learning, and policy development has had a significant effect on the United States and its communities. However, technology, economic development, and population growth continue, and two important questions cannot be answered: Are we doing enough quickly enough? Will we be able to give our children a sustainable and nonlethal world? In one respect, these questions have no answer, since none of us can predict the future. I would be satisfied with a trend line indicating that our ability to address environmental problems is keeping pace with the introduction of new problems.

Improvements Needed

Today, at the beginning of the twenty-first century, we must consider what is needed to improve environmental policy. A sample of such needs can be grouped into the following categories:

- Improved information about environmental conditions
- Better communication and understanding of that information
- Improved education of environmental professionals
- Creation and analysis of economic policies that lead to sustainable development

- Further advances in environmental analysis, pollution prevention, and mitigation capacity in government and industry
- Expanded development of community-based organizations and local institutions of government to operationally define and implement sustainable development strategies

Improved Information about Environmental Conditions

Currently our society still does not know enough about the status of our planet's environment. Economic data are more systematically collected than data about ecology. We need to establish a global observatory and take periodic systematic readings on environmental indicators. We also need to develop our modeling capacity to the point that we can project the impact of new technology and chemicals on human and ecological health. To do this, the U.S. government must invest substantial resources in developing the scientific knowledge and technical capacity to collect and analyze these measures. A global network should be created to connect government labs with universities, museums, and other organizations that can contribute to this effort. This network will cost a great deal of money but is essential to our survival.

Better Communication and Understanding of Environmental Data

Once this information is collected we need to analyze it, project its impact, and communicate it throughout the world. The political use of science in the climate change dispute should be seen as a warning. It took a very long time to develop a consensus on data that were far from ambiguous. Information is always used this way in politics. Economic data and program evaluation data are always subject to spin. We need to reduce this tendency as much as possible when we are interpreting environmental data. Like firefighters in a burning building, our lives may depend on this information, and so it needs to be objective and verifiable.

Those that collect baseline environmental data must make a special effort to simplify it, analyze its probable effect, and then communicate it to a wide audience. The environment is an issue with many complex scientific elements. It is not enough simply to collect and analyze data; the information must be presented in a form appropriate for policy decision making. Scientists who work with this information should not be expected to be able to communicate it to nonscientists. It is the task of communication experts and policy analysts to translate the information to policy makers and to the general public.

Improved Education of Environmental Professionals

The function of "translating" the information is not performed well today, as not enough people are sufficiently educated in science, policy making, and communication. More people need to be trained in fields such as environmental engineering, chemistry, climatology, toxicology, ecology, policy, finance, and management, to name only some of the disciplines that require greater emphasis. One way to educate people in these disciplines to enable them to work as practicing professionals is to place them in educational experiences that bring cross-disciplinary teams together to solve environmental problems.

We also need to develop a generalist environmental professional to assume a role in environmental protection similar to that played by an MBA in business or an MPA in the public sector. I have been involved over the past several years in an effort to adapt the Masters of Public Administration degree to a degree that would create a public-sector environmental professional. This new type of professional would be trained in science, communication, and the analysis of policy, politics, and management. The environmental problem is only solved by a thorough understanding of earth systems and the macro-level thinking required to manage those systems. This requires an understanding both of ecological and planetary sciences and of organizational and network management. Organizational management increasingly requires the coordination of networks of organizations rather than the command of vertically integrated hierarchies. The system-level thinking that is necessary for understanding ecological interactions is also needed to manage organizational networks devoted to environmental protection.

A profession is characterized by a shared view of how the world works or by a shared paradigm. Members of a profession have a common definition of problems and a standard set of skills to understand these problems and build solutions. Earth systems professionals must be able to integrate specific small-scale projects with a broader understanding of how the world works. René Dubos was correct when he identified the need to think globally and act locally. Earth systems professionals must also think systemically and act pragmatically. They must be able to move away from problems in the manner permitted by our incremental policy process, but at the same time they must seek to ensure that the solution to one problem is not the cause of another, more serious problem. Earth systems professionals must be able to develop innovative approaches to problem solving based on field-level empirical data with a profound understanding of earth systems functions and interactions.

The physical and natural sciences operate largely on different principles from the social and policy sciences. Although some social scientists have made an effort to imitate physical and natural science methodologies in their work, and these scientists share the worldview of natural and physical scientists, many social scientists and policy analysts see the world differently than scientists who study earth systems and ecology. Earth systems professionals must internalize both sets of norms and be able to translate from one to the other. They must be capable of understanding ecological science and be able to explain it to managers and policy makers. They must be able to explain policy processes, management, and politics to ecological and other physical and natural scientists. Their role is one of analyst, manager, translator, and interdisciplinary facilitator. The creation of this new profession is in its infancy, but its establishment is a prerequisite if we are to address environmental problems successfully.

Creation and Analysis of Economic Policies that Lead to Sustainable Development

The idea that ecological well-being is a necessary component of economic development, rather than an obstacle to economic growth, has not yet been accepted by the American public or its elected leaders. In addition to the term "sustainable development," these actions are representative of the environmental principle known as "ecological modernization." Ecological modernization states that environmental protection and conservation can be achieved by technological advances that yield increased efficiency. In order for this concept to gain wider currency, we need to demonstrate its realism and efficacy. A wide range of program initiatives from green architecture to conserve energy and save on utility bills to sustainable and less polluting agricultural practices need to be established, evaluated, and publicized.

The problem with the politics of sustainable development is that the benefits of ecologically sound economic development are societal in nature. An individual company may find it profitable to factor environmental issues into their business practices, but they may find that they can make more money in the short-term by ignoring the environment. Business profit-and-loss statements are reported quarterly, and corporate governance tends to be oriented toward short-term results and immediate gratification. Standard business practice allows for long-term investment through capital expenditures, and most business professionals understand that some investments take a while to pay off. The challenge is to develop organizational routines that foster investment in development that is ecologically sound and sustainable (Sachs 2005, 367). We have begun this process but still have a long way to go before environmentally sound business practices are commonplace.

Further Advances in Environmental Analysis, Pollution Prevention, and Mitigation Capacity in Government and Industry

The reason we need to test these sustainable development policy ideas and train more environmental professionals is that we need to increase organizational capacity to protect the environment. In 1994 the Bureau of Economic Analysis of the U.S. Department of Commerce estimated that $122 billion, or 1.76 percent of the gross domestic product (GDP), was spent on pollution abatement and control (U.S. Department of Commerce 1996). This analysis of environmental costs was discontinued after 1994. We certainly need to spend more than 1.76 percent of our gross domestic product on environmental protection and sustainable development, and in all likelihood we are spending much more than that today. In order to establish the appropriate funding level, it is necessary to conduct new analyses of how much of our GDP is being spent on environmental protection. Increased resource allocation to environmental protection is needed at all levels of government and in organizations that are not conventionally thought of as environmental organizations. Just as accountants are found in the financial divisions of most organizations and lawyers are found in the counsel office of agencies, a place must be found in every organization for professionals whose job is to ensure that best environmental management practices are implemented.

Government rules create the demand for these professionals. Without tax laws and the Security and Exchange Commission's regulation of financial markets, accountants and audited financial statements would not exist. The predictability and comparability of corporate behavior has created capital markets that are far larger and more dynamic than they would have been without regulation. The financial markets of the 1920s showed the potential of a mass capital market, and the crash of 1929 demonstrated the logical conclusion of "free" capitalism without the rule of law. The rules on capital markets and corporate financial reporting have not destroyed corporate capital finance—quite the contrary. Companies such as Enron that violate these rules are outliers that demonstrate the value of financial rules.

Can this be a model for the development of a class of environmental professionals in our private and public economy? Unlike capital finance, environmental protection is not a function that corporations see as vital to their fundamental health. The possibility of liability charges stemming from victim lawsuits provides a negative rationale for paying attention to environmental damage from corporate operations. On the positive side, there are considerable public relations benefits that stem from a "clean" environmental image. Finally, corporate ethics and citizenship could contribute to the development of environmentalism as an element of a corporation's culture.

This combination of regulation, fear of liability, public relations, and corporate ethics provides a compelling rationale for developing environmental professionalism. We need more of this capacity than we currently have in place. Although this group of professionals is growing rapidly, we have no way of knowing if the growth rate of this profession is fast enough to build a sustainable economy. The capacity we need will only be generated through the development of this group of professionals. The maintenance and expansion of environmental regulation is the only way, in my view, to ensure that this capacity is expanded. The greatest threats to the development of this capacity lie in the political extremes of our society. On the one hand, we have the very short-sighted anti-regulation fervor present in the American political right wing and in parts of corporate America. On the other, we have environmental extremists who consider all economic activity destructive and all corporations evil. Although some environmental rules have clearly been heavy-handed and counterproductive, and some corporations are indeed run by criminals, for the most part the rules work and private firms are not in business to poison people.

Our goal should be to develop and maintain a mature, sophisticated, and flexible environmental regulatory regime. This will result in the development of enhanced capacity to protect the environment.

Expanded Development of Community-Based Organizations and Local Institutions of Government to Define and Implement Sustainable Development Strategies

The behaviors that protect or damage the environment occur in specific local places. While many policy battles that influence local politics take place in Washington, without concurrence at the local level, policy decisions made in the capital are only partially if at all implemented. There seems to be a physical and political disconnect between environmental policies made in Washington and the places that will be affected by such policies. In 2004, for example, the decision to withdraw rules to protect areas without roads in parts of the wilderness in the western United States was advocated by governors who believed that their constituents favored reduced regulation. Pressure on those governors from the local level would be more important than action in Washington to change this decision and the political calculation it is based on. Even the Department of Energy's $60 billion dollar nuclear waste repository in Yucca Mountain, Nevada—clearly an issue of national importance—may be subjected to what must be seen as a local veto. In the United States, as I noted earlier, Tip O'Neill's famous truism that "all politics is local" means that the most important arena for improving environmental policy is local.

As Mazmanian and Kraft (1999) and Weber (2003) indicated, we need to pay attention to local organizations and the development of environmental institutions at the local level. Advocacy groups and the federal government in Washington, D.C., as well as their counterparts at the state level, play a role in environmental protection, but the behaviors they are trying to influence take place at the community level. If we are truly to institutionalize environmental protection, we need eyes and ears on the ground to keep track of the work done in environmental programs and the environmental impact of the work being done by government, private firms, and individuals.

The siting and expansion of businesses, transportation infrastructure, and other human made development initiatives occur at the local level, and their shape can best be influenced at that level. National rules are needed to ensure that communities cannot compete against one another by giving away pollution rights.

The development of institutional capacity at the local level will not be easy. Our society's increased environmental literacy should make it possible for environmental experts to be deployed at the local level, but the resources to pay for their expertise are likely to be scarce. Voluntary organizations can fill this gap by having environmental professionals help to assess local impacts on a voluntary basis in their home communities. Local voluntary organizations cycle in and out of life and perhaps cannot be depended on, but they do tend to emerge in response to specific development projects and can be an effective form of institution in some localities.

The alternative to voluntary organizations would be to create a special unit in county, city, or town governments. A local sustainable development agency of government could evolve out of the local unit that makes zoning decisions or issues building permits. This has already happened in certain places. An enlightened federal policy could provide funding for the first five to ten years of local environmental analysis and citizen participation. If the work done by this group proved essential and if it built a local constituency for its work, it could attract a long-term source of local funding. In fact, a local revenue stream taxing developers could be used to support this function over the long term.

Next Steps

The United States has been improving its environmental policy for more than thirty years, and an issue that did not exist in 1950 is now a potent political and economic force in our world. This chapter identified, in broad strokes, some of the accomplishments since the establishment of the U.S. EPA in 1970. Also discussed were some improvements we need in environmental policy making if we are to continue to make progress.

The complexity of the environmental problem can be daunting. It is easy to get discouraged and overwhelmed. We need to work on the environmental problems that come from our daily lifestyle, and we also need to think about eco-terrorism as political extremists seek to gain attention or harm their foes through the use of typically toxic weapons of mass destruction.

Predictions of the future are never easy and are often inaccurate. We do not know what new technologies will be invented or what type of evil force may achieve power in some nation. We do not understand the effects of our current actions on our environment, and so who would dare to predict events that have not even taken place?

When looking at the changes in policy making that I believe are essential, I confess that I search in vain for the motor—the mechanism to inspire people to devote more time and energy to protect the environment. In the past, great change has often followed great crises. Most of what we know about environmental change is that it is slow and steady; rarely is it rapid and dramatic. Worse still, most of the benefits of activities that damage the environment are immediate or short-term whereas the costs are largely long-term.

The key element missing in the equation in the United States has been presidential leadership. The closest we ever came to a president who understood the environment was Vice President Al Gore. For President Clinton, the overriding focus was "the economy, stupid!" For the two George Bushes, the issue of foreign policy dominated. Jimmy Carter understood the problem but had other priorities, and Nixon, Ford, and Reagan thought about the issue only in political terms. To these presidents the environment represented an interest group to mollify, not a national problem to address. For the environment to become a higher priority in the U.S., we need presidential leadership and the political courage that typically accompanies strong leadership. Because the current electoral process features a campaign finance system that is totally out of control, it is unlikely that a president in the near term will get out ahead of the moneyed interests that fund political campaigns. As long as our legal system considers donating funds to campaigns as a form of speech, the power of economic interests will dominate. One wonders whether the enlightened self-interest of the New Deal leadership of Franklin D. Roosevelt could ever happen in the current system.

While presidential leadership would accelerate the process, ultimately the environmental issue comes down to faith in humanity. Are we willing to do the work that is necessary to leave our children a world that has a reasonable possibility of surviving? John Kennedy's words, spoken at the American University in June 1963, with which I began this book merit repeating here, at the closing:

> For, in the final analysis, our most basic common link is that we all inhabit this small planet. We all breathe the same air. We all cherish our children's future. And we are all mortal.

Because we love our children and cherish their future, children are, in a way, a hedge against our inevitable demise, the recognition, in Kennedy's words, that "we are all mortal." But can our love for our children prevail over our own self-centered behavior? I like to think it can. I certainly hope it will. I suspect that the long-term survival of our species depends on it.

NOTE

1. This is given in constant dollars. In actual 1970 dollars the EPA budget was about $1 billion when the agency was founded.

References

Ackerman, B. A., and R. B. Stewart. 1988. "Reforming Environmental Law: The Democratic Case for Market Incentives." *Columbia Journal of Environmental Law,* 13:171–200.

Allen, T. 2004. Congressional Record 108th Congress. pp H1284, March 18.

Allison, G. T. 1971. *The Essence of Decision.* Boston: Little, Brown.

Allison, G. T., and P. Zelikow. 1999. *Essence of Decision: Explaining the Cuban Missile Crisis.* 2nd ed. New York: Longman.

American Automobile Association. 2004. "AAA Predicts Record July 4 Travel amid Strong Summer Travel Season." Retrieved on June 22, 2004 from http://www.aaanewsroom.net/Articles.asp?ArticleID=313&SectionID=4&CategoryID=8&SubCategoryID=&.

American Institute of Physics (AIP). 2003. *The Discovery of Global Warming: The Modern Warming Trend.* Retrieved July 7, 2004, from http://www.aip.org/history/climate/20ctrend.htm.

Bachrach, P., and M. S. Baratz. 1970. *Power and Poverty: Theory and Practice.* New York: Oxford University Press.

Barcott, B. 2004. "Changing All the Rules." *New York Times Magazine,* April 4.

Bardach, E., and R. A. Kagan. 1982. Introduction to *Social Regulation: Strategies for Reform,* ed. E. Bardach and R. A. Kagan. San Francisco: Institute for Contemporary Studies.

Barnett, H. C. 1994. *Toxic Debts and the Superfund Dilemma.* Chapel Hill: University of North Carolina Press.

Barnhardt, A., et al. 2004. *Mercury: Global Problems, Local Solutions.* A report for the Waterkeeper Alliance by graduate students in the MPA Program in Environmental Science and Policy at Columbia University, New York.

Bartsch, C., and E. Collaton. 1997. *Brownfields: Cleaning and Reusing Contaminated Properties*. Westport, Conn.: Praeger.

Betsill, M. M. "Global Climate Change Policy: Making Progress or Spinning Wheels?" 2004. In *The Global Environment: Institutions, Law, and Policy*, ed. R. S. Axelrod, D. L. Downie, and N. J. Vig, 2nd ed. Washington, D.C.: CQ Press.

Blackman, W. C. 2001. *Basic Hazardous Waste Management*. New York: Lewis.

Braybrooke, D., and C. Lindblom. 1963. *A Strategy of Decision*. New York: Free Press of Glencoe.

Breger, M. J., R. B. Stewart, E. D. Elliot, and D. Hawkins. 1991. "Providing Economic Incentives in Environmental Regulation." *Yale Journal on Regulation*, 8:463–495.

Breyer, S. 1982. *Regulation and Its Reform*. Cambridge, Mass.: Harvard University Press.

Broecker, W. S. 1975. "Climatic Change: Are We on the Brink of a Pronounced Global Warming?" *Science* 189, no. 4201: 460–464.

Brown, D. A. 2002. *American Heat: Ethical Problems with the United States' Response to Global Warming*. Boston: Rowman and Littlefield.

Bullard, R. D. 1992. "In Our Backyards: Minority Communities Get Most of the Dumps." *EPA Journal*, 18:11–12.

Caldwell, L. 1998a. *The National Environmental Policy Act: An Agenda for the Future*. Bloomington: Indiana University Press.

———. 1998b. "Witness Testimony at NEPA Hearing," March 11. Retrieved July 10, 2004, from http://resourcescommittee.house.gov/archives/105cong/fullcomm/98mar18/caldwell.htm.

Carson, R. 1962. *Silent Spring*. Boston: Houghton Mifflin.

Chicago Public Library. 2001. "Facts about Chicago." Retrieved June 6, 2005, from http://www.chipublib.org/004chicago/chifacts.html.

City of Chicago, Department of Streets and Sanitation. 2005. "Residential Garbage." Retrieved June 6, 2005, from www.egov.cityofchicago.gov.

Clapp, J. 2001. *Toxic Exports: The Transfer of Hazardous Wastes from Rich to Poor Countries*. Ithaca, N.Y.: Cornell University Press.

Cobb, R., and C. Elder. 1983. *Participation in American Politics: The Dynamics of Agenda Building*. Baltimore, Md.: Johns Hopkins University Press.

Cohen, J. E. 1995. *How Many People Can the Earth Support?* New York: W. W. Norton.

Cohen, S. 1984. "Defusing the Toxic Waste Time Bomb." In *Environmental Policy in the 1980s: Reagan's New Agenda*, ed. N. Vig and M. Kraft. Washington, D.C.: CQ Press.

Cohen, S "Employing Strategic Planning in Environmental Regulation." in Sheldon Kamieniecki, Richard Gonzales, and Robert Vos (eds.), *Flashpoints in Environmental Policymaking: Controversies in the 1990's*. Albany, N.Y.: SUNY Press, 1997.

Cohen, S., and R. Brand. 1993. *Total Quality Management in Government: A Practical Guide for the Real World.* San Francisco: Jossey-Bass.

Cohen, S., and W. Eimicke. 2002. *The Effective Public Manager: Achieving Success in a Changing Government.* San Francisco: Jossey-Bass.

Cohen, S., and D. Kamieniecki. 1991. *Environmental Regulation through Strategic Planning.* Boulder, Colo.: Westview.

Cohen, S., S. Kamieniecki, and M. A. Cahn. 2005. *Strategic Planning in Environmental Regulation.* Cambridge, Mass.: MIT Press.

Commoner, B. 1996 [1971]. "The Closing Circle: Nature, Man, and Technology." In *Thinking about the Environment: Readings on Politics, Property, and the Physical World,* ed. M. A. Cahn and R. O'Brien. Armonk, N.Y.: M.E. Sharpe.

Cornell Waste Management Institute. 2000. *Pay as You Throw for Large Municipalities.* New York City Pay-as-You-Throw Roundtable. Available at http://cwmi.css.cornell.edu/PAYTreport.pdf.

Davies, J. C., and J. Mazurek. 1998. *Pollution Control in the United States: Evaluating the System.* Washington, D.C.: Resources for the Future.

Day, C. 2004. "EU Climate Policy: Recent Progress and Outlook." Keynote address at the Second Brussels Climate Change Conference, May 11.

Deming, W. E. 2001. *Out of the Crisis.* Cambridge, Mass.: MIT Press.

Department of Sanitation, City of New York. 2000. "Comprehensive Solid Waste Management Plan, Draft Modification." New York: Author. Available at http://www.nyc.gov/htm/dos/pdf/pubnrpts/swmp-4oct.html.

———. 2001. *New York Recycling in Context: A Comprehensive Analysis of Recycling in Major U.S. Cities.* Available at http://www.nyc.gov/html/dos/html/recywprpts.htm/#3.

———. 2003. *Fact Sheet.* Retrieved June 15, 2004, from http://www.nyc.gov/html/dos/html/dosfact.html.

———. 2004. "Comprehensive Solid Waste Management Plan (Draft)." Retrieved August 29, 2005, from http://www.nyc.gov/html/dos/html/pubnrpts/swmp-4oct.html.

Earth Institute and the Urban Habitat Project. 2001. *Life after Fresh Kills: Moving beyond New York City's Current Waste Management Plan.* New York: Columbia Earth Institute, Columbia University.

Echeverria, J. D., and R. B. Eby. 1995. *Let the People Judge: Wise Use and the Private Property Rights Movement.* Washington, D.C.: Island Press.

Ecumenical Task Force of the Niagara Frontier (ETFNF). 1998. *Background on the Love Canal.* Retrieved April 28, 2004, from http://ublib.buffalo.edu/libraries/projects/lovecanal/background_lovecanal.html.

Edwards, S. M., T. C. Edwards, and C. B. Fields. 1996. *Environmental Crime and Criminality.* New York: Garland.

Feticiano, V. 1984. "Leaking Underground Storage Tanks: A Potential Environmental Problem." *Congressional Research Service Report,* S321–24, 1. Washington D.C.: Congressional Research Service.

Freeman, A. M., III. 1990. "Water Pollution Policy." In *Public Policies for Environmental Protection,* ed. P. Portney, 97–150. Washington, D.C.: Resources for the Future.

Gallup Organization. 2005a. "Daily Concerns Overshadow Environment Worries." Published April 19, 2005. Retrieved May 26, 2005, from www.gallup.com/poll/content/. Available at http://brain.gallup.com/content/default.aspx?ci=15925.

———. 2005b. "Half of Americans Say Bush Doing Poor Job on the Environment." Retrieved August 29, 2005, from http://brain.gallup.content/default.aspx?ci=1615.

———. 2005c. "Environment." Retrieved May 26, 2005, from www.gallup.com/poll/content/.

Gans, H. 1962. *Urban Villagers: Group and Class in the Life of Italian-Americans.* New York: Free Press.

Gibbs, L. 1998. *Learning from Love Canal: A 20th Century Retrospective.* Retrieved April 28, 2004, from http://arts.envirolink.org/arts_and_activism/LoisGibbs.html.

Hahn, R. W., and G. L. Hester. 1989. Where Did All the Markets Go? An Analysis of EPA's Emissions Trading Program. *Yale Journal on Regulation,* 6:109–153.

Hansen, J. E., et al. 1981. "Climate Impact of Increasing Atmospheric Carbon Dioxide." *Science* 213, no. 4511: 957–966.

Hawken, P., A. Lovins, and P. Lovins. 1999. *Natural Capitalism: Creating the Next Industrial Revolution.* Snowmass, Colo.: Rocky Mountain Institute.

Hird, J. A. 1994. *Superfund: The Political Economy of Environmental Risk.* Baltimore, Md.: Johns Hopkins University Press.

Holt, D. B., and J. B. Schorr. 2000. "Introduction: Do Americans Consume Too Much?" In *The Consumer Reader,* ed. J. B. Schorr and D. B. Holt, vii–xxiii. New York: New Press.

Information Unit on Climate Change (IUCC). 1993. *Intergovernmental Panel on Climate Change.* Retrieved January 18, 2005, from http://www.cs.ntu.edu.au/homepages/jmitroy/sid101/uncc/fs208.html.

Intergovernmental Panel on Climate Change (IPCC). 2001. "Summary for Policymakers." In *IPCC Third Assessment: Climate Change 2001.* Geneva: IPCC.

Jacobs, J. 1962. *The Death and Life of Great American Cities.* New York: Random House.

Jalonick, M. C. 2003. "Bush Signs 'Healthy Forests' Legislation into Law." *Congressional Quarterly Today,* December 3.

Johnston, J. L. 1991. A Market without Rights: Sulfur Dioxide Emissions Trading. *Regulation,* 14:24–29.

Jones. C. O. 1974. "Speculative Augmentation in Federal Air Pollution Policy-Making," *Journal of Politics* 36, no. 2: 438–463.

———. 1975. *Clean Air: The Policies and Politics of Pollution Control.* Pittsburgh: University of Pittsburgh Press.

Kolstad, C. 2000. *Environmental Economics.* New York: Oxford University Press.

Kraft, M., and R. Kraut. 1988. "Citizen Participation and Hazardous Waste Policy Implementation." In *Dimensions of Hazardous Waste Politics and Policy,* ed. C. E. Davis and J. P. Lester. New York: Policy Studies Organization.

Kraft, M., and N. J. Vig. 1984. "Environmental Policy from the 70s to the 80s." In *Environmental Policy in the 1980s: Reagan's New Agenda,* ed. N. Vig and M. Kraft, 3–26. Washington, D.C.: CQ Press.

———. 1990. "Environmental Policy from the Seventies to the Nineties: Continuity and Change." In *Environmental Policy in the 1990s: Toward a New Agenda,* ed. N. Vig and M. Kraft. Washington, D.C.: CQ Press.

Lackner, C. S. 2003. "Climate Change: A Guide to CO_2 Sequestration." *Science* 300:1677–1678.

Landsberg, H. E. 1970. "Man-made Climatic Changes." *Science* 170, no. 3964: 1265–1274.

Landy, M. K., M. J. Roberts, and S. R. Thomas. 1990. *The Environmental Protection Agency: Asking the Wrong Questions.* New York: Oxford University Press.

Layzer, J. 2002. *The Environmental Case: Translating Values into Policy.* Washington, D.C.: CQ Press.

Leopold, A. 1949. *A Sand County Almanac, and Sketches Here and There.* New York: Oxford University Press.

Levenson, L. and D. Gordon. 1990. Drive+: Promoting Cleaner and More Fuel Efficient Motor Vehicles through a Self-financing System of State Sales Tax Incentives. *Journal of Policy Analysis and Management,* 9:409–415.

Levine, A. G. 1982. *Love Canal: Science, Politics, and People.* Lexington, Mass.: D.C. Health.

Levy, D. L., and P. J. Newell, eds. 2004. *The Business of Global Environmental Governance.* Cambridge, Mass.: MIT Press.

Lewis, M. 2003. "McCain–Lieberman 2003: Dangerous Congressional Emissions on Global Warming." *National Review Online,* October 29.

Litan, R. E. and W. D. Nordhaus. 1983. *Reforming Federal Regulation.* New Haven: Yale University Press.

Lowry, R. C. 1998. "All Hazardous Waste Politics Is Local: Grass-roots Advocacy and Public Participation in Siting and Cleanup Decisions." *Policy Studies Journal* 26, no.4: 748–759.

Mazmanian, D., and M. E. Kraft. 1999. "The Three Epochs of the Environmental Movement." In *Towards Sustainable Communities: Transition and Transformation in Environmental Policy,* ed. D. Mazmanian and M. E. Kraft. Cambridge, Mass.: MIT Press.

Mazmanian, D., and D. Morell. 1988. The Elusive Pursuit of Toxics Management. *Public Interest,* 90:81–98.

———. 1990. "The NIMBY Syndrome: Facility Siting and the Failure of Democratic Discourse." In *Environmental Policy in the 1990s: Toward a New Agenda*, ed. N. Vig and M. Kraft. Washington, D.C.: CQ Press.

McCrory, J. 1998. "New York City: The First Regional Government Still Cries for Planning the Case of Waste Management." *Planners Network*, no. 128 (March/April). Retrieved August 29, 2005, from http://www.plannersnetwork.org/htm/publications/1998_128/McCrory.htm.

Meir, K. J. 1985. *Regulation: Politics, Bureaucracy and Economics*. New York: St. Martin's.

Milbrath, L. 1984. *Environmentalists: Vanguard for a New Society*. Albany: State University of New York Press.

Miller, B. 2000. *Fat of the Land: Garbage in New York, the Last Two Hundred Years*. New York: Four Walls Eight Windows; distributed by Publishers Group West.

Mitchell, R. C. 1984. "Public Opinion and Environmental Politics in the 1970s and 1980s." In *Environmental Policy in the 1980s*, ed. N. Vig and M. Kraft, 51–74. Washington, D.C.: CQ Press.

Mohai, P., and B. Bryant. 1998. "Is There a "Race" Effect on Concern for Environmental Quality?" *Public Opinion Quarterly* 62, no. 4: 475–505.

Moroney, J. R. 1998. "Energy, Carbon Dioxide Emissions, and Economic Growth." Washington, D.C.: American Council for Capital Formation. Retrieved June 14, 2005, from http://accf.org/publications/reports/sr-energy-carbon-growth.html.

Nakamura, R. T., and T. W. Church. 2003. *Taming Regulation: Superfund and the Challenge of Regulatory Reform*. Washington, D.C.: Brookings Institution.

National Solid Wastes Management Association. 2004. Retrieved June 2, 2005, fromhttp://wastec.isproductions.net/webmodules/webarticles/anmviewer.asp?a = 459&z = 44.

Odum, H. T., and E. C. Odum. 1981. *Energy Basis for Man and Nature*. New York: McGraw Hill.

Office of Management and Budget, Executive Office of the President. 2003. *Informing Regulatory Decisions: 2003 Report to Congress on the Costs and Benefits of Federal Regulations and Unfunded Mandates*. Washington, D.C.: OMB.

Office of New York State Attorney General Eliot Spitzer. 2003. "Watershed Study Cites Opportunity to Improve NYC Drinking Water and Catskill Waterway." Retrieved from www.oag.state.ny.us/press/2003.

O'Leary, R., R. Durant, D. Fiorino, and P. S. Weiland. 1999. *Managing for the Environment: Understanding Legal, Organizational, and Policy Challenges*. San Francisco: Jossey-Bass.

Ophuls, W., and A. S. Boyan. 1992. *Ecology and the Politics of Scarcity Revisited: The Unraveling of the American Dream*. New York: W. H. Freeman.

Pedersen, W. F., Jr. 1991. "The Future of Federal Solid Waste Regulation." *Columbia Journal of Environmental Law*, 16:109–142.

Perlman, H., W. B. Solley, and R. R. Pierce. 1985. *Estimated Use of Water in the United States in 1995* (U.S. Geological Survey Circular 1200). Washington, D.C.: U.S. Government Printing Office. Available at http://water.usgs.gov/wateruse/pdf1995/html/.

Pianin, E. 2003. "Superfund to Run Out of Money, GAO says." *Washington Post,* September 3, A15.

Pressman, J., and A. Wildavsky. 1984. *Implementation*. 3rd. exp. ed. Berkeley: University of California Press.

Probst, K., D. Konisky, R. Hersh, M. B. Batz, and K. D. Walker. 2001. *Superfund's Future: What Will It Cost?* Washington, D.C.: Resources for the Future.

Rabe, B. G. 2000. "Environmental Policy from the 70s to the Twenty-first Century." In *Environmental Policy: New Directions for the 21st Century,* ed. N. Vig and M. Kraft Washington, D.C.: CQ Press.

Reisch, M., and D. Bearden. 1997. *Superfund Fact Book*. Congressional Research Service Report 97–312. Washington D.C.: Congressional Research Service. Distributed by the National Library for the Environment at http://www.ncseonline.org/NLE/.

Rosenbaum, W. A. 2005. *Environmental Politics and Policy*. 6th ed. Washington, D.C.: CQ Press.

Rubin, C. T., ed. 2000. *Conservation Reconsidered: Nature, Virtue, and American Liberal Democracy*. Lanham, Md.: Rowman and Littlefield.

Sabatier, P. A., and H. C. Jenkins-Smith. 1993. *Policy Change and Learning: An Advocacy Coalition Approach*. Boulder, Colo.: Westview.

Sachs, J. 2005. *The End of Poverty*. New York: Penguin.

San Francisco Department of the Environment. 2005. Retrieved June 2, 2005, from http://www.sfenvironment.com/aboutus/recycling.

Schneider, A.L. and H. Ingram. 1997. *Policy Design for Democracy*. Lawrence: University Press of Kansas.

Seidman, H. 1986. *Politics, Position, and Power: The Dynamics of Federal Organization*. London: Oxford University Press.

Stroup, R. L. and J. S. Shaw. 1989. "The Free Market and the Environment." *Public Interest,* 97:30–43.

Stout, David. 2005. "Senate Panel Backs EPA Nominee, with One Notable Exception." *New York Times,* April 13.

Sunstein, C. R. 1990. "Paradoxes of the Regulatory State." *University of Chicago Law Review,* 57:407–441.

Swiss Re. 1998. *Climate Research Does Not Remove the Uncertainty: Coping with Risks of Climate Change*. Retrieved June 7, 2004, from http://www.swissre.com.

———. 2003. "Swiss Re Implements Ten-Year Program to Become Fully Greenhouse Neutral." Corporate press release, October 30. Retrieved June 11, 2003, from http://www.swissre.com.

Switzer, J. V., and G. Bryner. 1998. *Environmental Politics: Domestic and Global Dimensions,* 2nd ed. New York: St. Martin's.

Thompson, W. C. 2004. "No Room to Move: New York City's Impending Solid Waste Crisis." City of New York, Office of the Comptroller.

Tripp, J. T. B., and D. J. Dudek. 1989. "Institutional Guidelines for Designing Successful Transferable Rights Programs." *Yale Journal on Regulation,* 6:369–390.

Twitchell, J. 2000. "Two Cheers for Materialism." In *The Consumer Reader,* ed. J. B. Schorr and D. B. Holt, 281–290. New York: New Press.

U.S. Census Bureau. 2002. *Population Profile of the United States (2000).* Retrieved June 22, 2004, from http://www.census.gov/population/pop-profile/2000/chap03.pdf.

———. 2003. *Homeownership Rates by Citizenship, 1994 to 2002.* Retrieved July 6, 2004, from http://www.census.gov/hhes/www/housing/movingtoamerica2002/tab1.html.

U.S. Department of Commerce, Bureau of Economic Analysis. 1996. *Pollution Abatement and Control Expenditures, 1972–1994.* Washington, D.C.: Department of Commerce.

———. 1994. "Summary of National Income and Product Series, 1929–1993." Retrieved from http://www.bea.gov/bea/articles/NATIONAL/NIPA/1994/0994nip3.pdf.

U.S. Environmental Protection Agency (EPA). 1985. Office of Water. *Leaking Underground Storage Tanks.* Washington, D.C.: Author.

———. 1992. Superfund Program. "Progress toward Implementing Superfund, Fiscal Year 1992." Retrieved June 10, 2004, from http://www.epa.gov/superfund/accomp/sarc/1992/cvr92.pdf.

———. 1996. National Environmental Education Advisory Council. "Report Assessing Environmental Education in the U.S. and the Impact of the National Environmental Education Act of 1990." Retrieved May 6, 2005, from www.epa.gov/enviroed/pdf/preport.pdf.

———. 1998. "Twenty-five Years of Environmental Progress at a Glance." Retrieved May 8, 2004, from http://www.epa.gov/history/topics/25year/INTRO.PDF.

———. 1999. *Oil Program Update* 2, no. 2: 3.

———. 2000. *Global Warming–Climate Change.* Retrieved August 29, 2005, from http://yosemite.epa.gov/oar/globalwarming.nsf/content/climate.html.

———. 2001. Office of Emergency and Remedial Response. *The Mega Site Issue.* Retrieved June 9, 2005, from www.epa.gov/oswer/docs/naceptdocs/the_megasite_issue_final_61902.pdf.

———. 2002a. Office of Solid Waste. "Basic Facts: Municipal Solid Waste." Retrieved September 15, 2004, from www.epa.gov/eap/oswer/non-hw/muncpl/facts.htm.

———. 2002b. Office of Underground Storage Tanks. *How Have the UST Requirements Helped Protect the Environment?* Retrieved August 21, 2002, from *www.epa.gov/oust/faqs/topfour.html.*

———. 2002c. Office of Underground Storage Tanks. *Cleaning up UST System Releases*. Retrieved September 6, 2005, from www.epa.gov/OUST/cat/index.html.

———. 2002d. *25 Years of RCRA: Building on Our Past to Protect our Future*. Retrieved April 28, 2004, from http://www.epa.gov/epaoswer/general/k02027.pdf.

———. 2003a. Office of Water. "Overview of the National Water Program." Retrieved June 14, 2004, from http://www.epa.gov/water/programs/owintro.html.

———. 2003b. *Superfund Accomplishment Figures, FY 2003*. Retrieved April 28, 2004, from http://www.epa.gov/superfund/action/process/numbers.htm.

———. 2003c. *Superfund for Students and Teachers: Common Cleanup Methods*. Retrieved June 30, 2004, from www.epa.gov/superfund/students/clas_act/haz-ed/ff_08.htm.

———. 2004a. Office of Solid Waste. "Municipal Solid Waste: Basic Facts." Retrieved June 10, 2004, from www.epa.gov/epaoswer/non-hw/muncpl/facts.htm.

———. 2004b. *Preliminary Assessment/Site Inspection*. Retrieved July 7, 2004, from http://www.epa.gov/superfund/whatissf/sfproces/pasi.htm.

———. 2005a. Office of Underground Storage Tanks. *UST (Underground Storage Tank) Program Facts*. Retrieved June 7, 2005, from http://www.epa.gov/swerust1/pubs/ustfacts.pdf.

———. 2005b. *Superfund Budget History*. Retrieved June 21, 2005, from http://www.epa.gov/superfund/action/process/budgethistory.htm.

U.S. Geological Survey (USGS). 2005. "Estimated Use of Water in the United States in 2000." Table 14: Trends in Estimated Water Use in the United States, 1950–2000. Retrieved September 9, 2005, at http://pubs.usgs.gov/circ/2004/circ1268/htdocs/table14.html.

U.S. Government Accounting Office (GAO). 1999. *Superfund: Progress Made by EPA and Other Federal Agencies to Resolve Program Management Issues*. Washington, D.C.: U.S. GAO.

———. 2003. *Superfund Program: Current Status and Future Fiscal Challenges*. Washington, D.C.: U.S. GAO.

United Nations Framework Convention on Climate Change (UNFCCC). 2002. *A Guide to the Climate Change Convention and Its Kyoto Protocol*. Bonn: UNFCCC Climate Change Secretariat.

Victor, D. G. 2004. *Climate Change: Debating America's Policy Options*. New York: Council on Foreign Relations.

Weber, E. P. 2003. *Bringing Society Back In: Grassroots Ecosystems Management, Accountability, and Sustainable Communities*. Cambridge, Mass.: MIT Press.

White House. 2002. *Global Climate Change Policy Book.* Retrieved July 7, 2004, from http://www.whitehouse.gov/news/releases/2002/02/climatechange.html.

Wiedenbaum, M. 1992. Return of the 'R' Word: The Regulatory Assault on the Economy." *Policy Review,* 59:40–43.

Index